Tourism in European Microstates and Dependencies

Geopolitics, Scale and Resource Limitations

Tourism in European Microstates and Dependencies

Geopolitics, Scale and Resource Limitations

By

Dallen J. Timothy
Professor
Arizona State University

CABI

CABI is a trading name of CAB International

CABI
Nosworthy Way
Wallingford
Oxfordshire OX10 8DE
UK

Tel: +44 (0)1491 832111
Fax: +44 (0)1491 833508
E-mail: info@cabi.org
Website: www.cabi.org

CABI
WeWork
One Lincoln St
24th Floor
Boston, MA 02111
USA

T: +1 (617)682-9015
E-mail: cabi-nao@cabi.org

A catalogue record for this book is available from the British Library, London, UK.

Library of Congress Cataloging-in-Publication Data

Names: Timothy, Dallen J., author.
Title: Tourism in European microstates and dependencies : geopolitics, scale and resource limitations / by Dallen J. Timothy.
Description: Wallingford, Oxfordshire, UK ; Boston, MA, USA : CAB International, 2020. | Includes bibliographical references and index. | Summary: "Tourism in European Microstates and Dependencies carefully examines the nuances and realities associated with tourism, social and economic development, geography, and geopolitics of Europe's microstates and dependences"-- Provided by publisher.
Identifiers: LCCN 2020026191 (print) | LCCN 2020026192 (ebook) | ISBN 9781789243109 (hardback) | ISBN 9781789243116 (ebook) | ISBN 9781789243123 (epub)
Subjects: LCSH: Tourism--Europe. | States, Small--Europe. | Geopolitics--Economic aspects--Europe.
Classification: LCC G155.E8 T56 2020 (print) | LCC G155.E8 (ebook) | DDC 338.4/7914--dc23
LC record available at https://lccn.loc.gov/2020026191
LC ebook record available at https://lccn.loc.gov/2020026192
References to Internet websites (URLs) were accurate at the time of writing.

ISBN-13: 9781789243109 (hardback)
9781789243116 (ePDF)
9781789243123 (ePub)

Commissioning Editor: Claire Parfitt
Editorial Assistant: Lauren Davies
Production Editor: James Bishop

Typeset by Exeter Premedia Services Pvt Ltd, Chennai, India
Printed and bound in the UK by Severn, Gloucseter

Table of Contents

List of Figures

List of Tables

Acknowledgements

This book would not be possible without the vital assistance of hundreds of mayors, town officials, governors, border officers, local residents, destination management organization representatives, tourism service providers, librarians, and tourists, who were part of my fieldwork and provided volumes of information about local life and tourism in Europe's unique territories between 1995 and 2019. Such an endeavor would also not have been possible without amazing friends and colleagues who tolerated my border hopping and accompanied me on multiple visits to enclaves, microstates, dependencies, de facto states, micronations, and autonomous areas. Special recognition goes to my dear friend and colleague, Professor Jaume Guia, for organizing so many fruitful research trips to the microstates, exclaves and micronations of Southern Europe. It has been a pleasure to learn from you and with you. I owe a huge debt of gratitude to His Eminence, Cardinal Giovanni Lajolo, former President of the Governorate of Vatican City State, for permission to explore, learn and inquire as a guest of the state. To my colleague and friend, Dr. Alon Gelbman, I am grateful for your willingness to accompany me gleefully and enthusiastically on so many border and exclave research trips during the past 15 years!

Other colleagues, friends and family members who have contributed in important ways to the development of this work include: Dani Blasco, Judit Diaz, Nico Berthet, Lluís Prats, Lluís Mundet, Núria Galí, Tomas Cuevas Contreras, Isabel Zizaldra Hernandez, Teresa Martínez Almanza, Delann and Kelly Denham, Noga Collins-Kreiner, Amos Ron, Milan Topic, Dimitri Ioannides, Peter Dervin, Stephen Boyd, Alan Lew, Myriam Jansen-Verbeke, Anne Soper, Jiayi (Kerry) Lin, Tim Coles, Brendan Whyte, Hugh Wallis, and Terhi and Juhani Kirstilä. I extend a special thank you to my wife, Carol, and my daughter, Olivia, for their never-ending patience as they accompanied me on several fieldwork trips to Europe, and for my sons Kendall, Aaron and Spencer, who patiently remained at home.

1

Europe's Smallest States and Anomalous Geographical Areas

Introduction

To state the obvious, the world is comprised of many places, yet every place is different. The term 'place' entails everything from large continents down to microscale localities of individual importance and personal spheres of influence (e.g. home, garden, and workspace). Between these two scalar extremes, the land surface of the earth has evolved through human history into various governed entities, including supranational regions, individual sovereign states, and subnational administrative units such as provinces, states, cantons, oblasts, emirates, counties, parishes, districts, municipalities, and neighborhoods. Each country has its own way of delegating power and authority to subnational polities and ascribing certain administrative and service responsibilities to them. The degree to which nations or local political units are empowered depends somewhat on whether a country is a unitary state or a federated state, or a democracy or an absolute monarchy. Thus, the nature of the state matters in international affairs and power relations in domestic governance.

Although volumes could be written about the geopolitics of tourism at any of these levels—supranational, national, or subnational, this book focuses on the sovereign state level of governance and political geography. It is concerned with how the state manifests in the ordinary lives of people and places, with a particular emphasis on tourism.

This book is foremost about tourism and its multidimensional relationships with some aspects of state geopolitics as manifested on the ground in mini microstates and other anomalous territories in Europe. What they all have in common is their geographical and geopolitical marginality and separateness, their small physical sizes, and their relatively small populations. These characteristics differentiate them from the normative perspective of sovereign states, which receive almost all research and policy attention throughout the world. This introductory chapter provides an initial conceptual grounding for studying irregular territories and microstates and establishes definitions and terms that will be used all through the book, including sovereignty, state borders, state sizes, and the idea that many countries have geographical outliers that do not fit the standard definition of state, region, territory or even tourist destination.

Sovereignty, the State and International Borders

According to international law, sovereignty is the absolute right of an acknowledged entity to govern over itself and its affairs without impediments from outside sources or external bodies. It usually refers to ultimate authority over a physical territory or polity, in most cases a state and its possessions. A sovereign state possesses the legal right to legislate and regulate in all areas of social, political, economic, legal and ecological matters within, and extending up to, its national borders (Dahlman, 2009a; Glassner and Fahrer, 2004; Timothy, 2001c).

Size has never truly been a criterion for statehood, although as shown later, it has been an impediment. Instead, to be a *sovereign state*, a country must possess four characteristics (Connell, 1991; Glassner and Fahrer, 2004). First, it must have physical territory. To be legally recognized in accordance with Westphalian sovereignty (the basis of current international law), a state must be comprised of physical land. There is today only one sovereign entity (organization) that is not a sovereign state: the Sovereign Military Order of Malta, which will be discussed in greater detail in Chapter 9. Historically, governments in exile, such as those of the Netherlands and Luxembourg during the Second World War, were considered sovereign entities while their territories were occupied by Germany. Second, a sovereign state must have a permanent population. There is no stipulation about population size requirements for sovereignty. Presently, demographic sizes of sovereign states vary from 1.4 billion to less than 1000. Third, a functioning government system is required. A government, in any form, must administer over the people and territory of the state, providing protection and public services (e.g. postal services, utilities, transportation networks, water and utilities), and overseeing a functioning economy. Finally, a country can be regarded as sovereign only if it has the capacity to negotiate directly with other sovereign states. This typically means the power to enter into treaties with other countries or supranational organizations, sign trade agreements, and establish diplomatic relationships. This also includes categorical recognition by a significant majority of existing states (Glassner and Fahrer, 2004).

All sovereign states are independent countries, but not all countries are sovereign states. For example, Aruba, Curaçao, Sint Maarten, and the Netherlands are the four 'constituent countries' that comprise the sovereign state of the Kingdom of the Netherlands. Similarly, England, Scotland, Wales, and Northern Ireland are the four 'countries' that make up the United Kingdom (UK), which is a sovereign state, but the four constituent parts of the UK are not sovereign states. The same situation is true of the Kingdom of Denmark, which comprises the constituent countries of Denmark, Greenland and the Faroe Islands. The eleven countries mentioned here do not possess sovereignty, but they do have a high level of autonomy within the states to which they belong.

A *nation* is not a state or country but a people who share a common culture, mutual values, history, and a homeland. For example, Native American tribes are often referred to as nations. Albanians, in Albania and Kosovo, are the Albanian nation. Nationalism, or a sense of belonging to a specific nation, does not always correspond with the present boundaries of a state. The notion of a broader nation beyond state borders often gives way to the growth of *irredentism*, or efforts by State A to incorporate certain areas of States B or C that are populated by the same nations and which might have been part of State A at some point in history (Dahlman, 2009b). The idea of a nation without its own state also can result in other situations such as *secessionism*—the desire of a 'nation' to break away from the state where it is currently located in an effort to establish its own state or country, or join another one. Prominent current situations of this in Europe include Catalonia's fervent efforts to secede from Spain and Scotland's efforts to leave the UK. Nationalism, irredentism and secession are key concepts in understanding some of the anomalous territories that will be highlighted later in the book.

A *nation-state* is a state that is also a nation (Gilmartin, 2009). It is a nation with its own state, where 'there is not a significant group that is not part of the nation. This does not mean simply a minority ethnic group, but a nationalistic group that either wants its own State or wants to be part of another State' (Glassner and Fahrer, 2004, p. 34). Examples of nation-states include Armenia, Mongolia and San Marino, each of which is populated at least 95% by a single nation. The terms 'sovereign state', 'independent country' (or state and country), and 'nation' are used throughout this book in their appropriate contexts, although state and country will be used interchangeably in certain places.

Borders, or boundaries (also sometimes referred to as frontiers, although this has a different historical connotation), are the physical or symbolic membranes that enclose the state and its rights of sovereignty. They have no width or depth, for they are simply the vertical planes where sovereign polities meet. Borders are the limits up to which a state can exercise control over its own dealings. Beyond its borders, a state has no authority.

Political scholars have identified several different types of borders, depending on their history and their composition (Hartshorn, 1936; Jones, 1959). First, are 'antecedent' (or pioneer) boundaries, which are formed in areas prior to human settlement and land use. In this case, settlement patterns, land tenure, and physical development are shaped, constrained, and maneuvered by the international boundary. Architectural styles, languages, religions and agricultural patterns frequently differ on opposite sides of antecedent borders. The second type is 'subsequent', or 'consequent', borders, which are drawn based upon emerging patterns of human settlement, as different groups (perhaps nations) become identifiable. In this case, clear human boundaries emerge, which then become political boundaries. Many borders are formulated after a region has already been settled and developed (Gelbman and Timothy, 2019; Newman, 2006). In this case, borders may be 'superimposed' or drawn regardless of what already exists in their paths. Superimposed borders frequently divide ethnic groups and nations, as well as towns, villages, farm fields, buildings, and even people's individual homes. Finally, 'relict' boundaries have been decommissioned. They no longer function as political borders, but they are still somehow visible in the cultural landscape (Paasi, 2013; Timothy, 2001c). A common example of this in the realm of heritage and tourism is the former Berlin Wall and entire East-West Germany divide, which are still marked by an embedded trail in Berlin along the former course of the wall and along the former Iron Curtain with several border museums, and wall pieces still standing (Gelbman and Timothy, 2010).

The normative process of border creation has been examined throughout the world in many different contexts (Klemenčić and Gosar, 2000; Thao, 2000; Yoon, 2014). According to Jones (1945, see also Donaldson and Williams, 2008), the ideal process of boundary formation takes place in four phases. First, borders are 'defined', meaning that they are described in detail vis-à-vis the geography and topography they traverse. This usually entails using natural or cultural features, such as streams, mountain summits, large boulders, or roads and fences, to describe the boundary line as precisely as possible. Second, borders are 'delimited', or plotted on maps, aerial photographs, and satellite images. This enables the frontier to be charted precisely. The third step is 'demarcation', which involves teams working in the field with survey equipment and means or tools for physically marking the border. These might include concrete obelisks or pillars, cut stones, metal posts, or fiberglass signs. This is the stage in which the precise borderline is physically laid out on the ground, indicating the exact vertical plane to which each country's sovereignty extends. Some boundary stones have become important historical markers and tourist attractions in their own right. Finally, in most cases, border demarcation is periodically checked and monitored to ensure that natural processes or intentional human actions have not altered the location of the boundary.

Along the French-Spanish border in the Cerdanya Valley, the Spanish Civil Guard (*Guardia Civil*) trek the course of the border each year to check the condition of boundary demarcation stones (erected in 1868) and how natural processes (i.e. plant growth or erosion) might have obstructed the view of the borderline. This has become somewhat of a leisure pursuit for local residents and tourists on the Spanish side of the border who sometimes accompany the officials on their mountain treks. It is important to note that this four-stage process is an ideal and under 'normal' circumstances; not all borders were created in such a legal and systematic way. Many were formed through acts of aggression or as remains of feudal property ownership during medieval times, and in other ways.

Besides indicating the precise limits of a country's sovereign rule, borders have other functions as well. In many instances, they serve as lines of military and security defense. Many contemporary borders were the locations of former battlefronts and ceasefire lines. Some are still frontlines in active combat zones. Others are frontlines in 'wars' on drugs and terrorism. Secondly, boundaries are filters. They prevent the importation of illegal substances, potentially contaminated food items, laundered money, or endangered species. Relatedly, the third role is a means of protecting domestic industries and economic development by levying import duties and tariffs, and setting limits on the numbers of certain items that can be imported by individuals or companies. Fourth, they are a tool for monitoring or hindering human mobilities. They are legal impediments for people wishing to enter a country illegally. They may manifest as strong physical barriers with fences, walls and minefields, or administrative barriers that are difficult to cross without adequate documentation. Passports and visas are checked at border crossings for people's identities and eligibility to enter.

State boundaries have been recognized for a long time as important tourism venues. Borders function as tourist attractions in several different ways (Timothy, 1995, 2000, 2001c; Timothy *et al.*, 2016; Więckowski, 2010). First, they attract curiosity seekers when they manifest in unique and anomalous ways in the cultural landscape (e.g. ornate gateways and famous walls). Second, borderlines often divide ideological regimes. Opposing political systems meeting at a specific line have drawn tourists' attention for decades, and observation platforms are often built so onlookers can see what lies in the 'forbidden other'. Third, many national frontiers divide specific attractions, such as buildings, golf courses, or waterfalls. Fourth, borders routinely divide beaches, mountain peaks, or even archaeological sites, which can add a certain level of fascination to an already interesting place. Fifth, border-themed attractions have been built in many localities to commemorate the border heritage of a region. These include border museums, peace parks, and camping/picnic areas. Finally, borders and their demarcations are often treated as historic markers that are interpreted with signage and linked together as heritage trails for hikers, cyclists or car-based travelers.

Beyond the line itself, frontier areas in some countries are popular destinations because their border locations enable them to establish certain activities that might not be permitted on the opposite side (e.g. gambling and prostitution), because services are much cheaper (e.g. medical tourism) or because lower taxes render certain consumer items much less expensive than at home (e.g. pharmaceuticals and utilitarian shopping) (Timothy *et al.*, 2016).

The second relationship is borders as barriers to tourism (Gelbman, 2010; Timothy, 2001c). As noted previously, state boundaries frequently hinder individual travel. Borders may be physical barriers that are meant to prevent people from crossing or psychological barriers when people are fearful of crossing because of cultural, political and economic differences on the other side or when currency exchange rates are unfavorable (Timothy, 2001c).

Third, tourism landscapes and borderscapes work together to create unique border tourismscapes that determine urban morphology in most border communities. Borderland tourism also creates distinct servicescapes (e.g. certain border-related business and public services) that do not exist in non-frontier areas. This has important implications for urban planning and tourism planning in borderlands (Prokkola, 2008) and often calls for increased transfrontier cooperation (Jakola and Prokkola, 2018). Finally, and perhaps the most routine relationship of all is borders as transit spaces. Although this relationship has important consequences for border tourismscapes as certain transit services are needed, most of the time, boundaries are crossed as quickly and effortlessly as possible, as travelers hasten to someplace more appealing (Timothy, 2001c).

Sizes of States

Beginning in earnest in the 1950s and 1960s (Robinson, 1960), there is now a protracted and robust discussion in the fields of geopolitics and law about the size of states, how size is defined, and how it affects governance (Baldacchino and Wivel, 2020; Dumieński, 2014). Early perspectives on size referred primarily to a country's physical dimensions. Later discussions introduced other variables that also contribute to understanding a state's absolute and relative size. Geographical area, population, level of development and GNP, degree of autonomy, power relations vis-à-vis larger states and supranational alliances, small military capacities, the number of diplomatic missions, and how a state perceives itself and how other states perceive it have all been used as proxy measures for state size (Baldacchino, 2018; Baldacchino and Wivel, 2020; Dommen and Hein, 1987; Dumieński, 2014; Goetschel, 1998; Hoffmann, 2016; Liu and Jenkins, 1996; Steinmetz and Wivel, 2010; Thorhallsson, 2018, 2019; Vital, 2006).

The physical size of a country has long been seen as either an advantage or a disadvantage to socio-economic development and good governance. The idea that small states are at a development disadvantage compared to physically larger states has been forwarded since the mid-1900s (Robinson, 1960), but there is no universally accepted definition of what constitutes a small state or a microstate, despite many efforts to determine this. Glassner and Fahrer (2004, p. 64) suggest that very large states exceed 2.5 million km², whereas those under 25,000 km² are very small states, with a range of medium-large and medium-small countries in between. Glassner and Fahrer (2004) propose that countries with areas of 5000 km² to 10,000 km² (e.g. Cyprus, The Gambia, Brunei) are mini states (Table 1.1), while the very smallest countries (e.g. Monaco, Liechtenstein, San Marino) are microstates. In the past, terms such as 'ministates', 'Lilliputian states' and 'very small states' have been used in official descriptions of the smallest microstates (Dumieński, 2014; Stringer, 2013). Researchers have at times even defined Iceland, Singapore, Finland, and Switzerland as small states or microstates (Dumieński, 2014; Guo and Wu, 2016; Ingebritsen *et al.*, 2006; Kelman and Shreve, 2015) based on economic growth, international relationships and population or a mix of variables, yet they are physically and demographically much larger than the smallest microstates of Europe.

Table 1.1. The world's smallest states by physical size

Position	Sovereign state	Total size in area (km²)
1	Vatican City	0.44
2	Monaco	2
3	Nauru	21
4	Tuvalu	26
5	San Marino	61
6	Liechtenstein	160
7	Marshall Islands	181
8	St Kitts and Nevis	261
9	Maldives	298
10	Malta	316
11	Grenada	344
12	St Vincent & the Grenadines	389
13	Barbados	430
14	Antigua and Barbuda	443
15	Seychelles	455
16	Palau	459
17	Andorra	468
18	St Lucia	616
19	Federated States of Micronesia	702
20	Singapore	719
21	Tonga	747
22	Dominica	751
23	Bahrain	760
24	Kiribati	811
25	São Tomé and Príncipe	964
26	Mauritius	2040
27	Comoros	2235
28	Luxembourg	2586
29	Samoa	2831
30	Cape Verde	4033

Source: Central Intelligence Agency (2020)

From a demographic viewpoint, smallness is also relative, although most scholars concede that a standard demographic measure of microstatehood should be a population of less than a million (Baldacchino, 1993; Connell, 1991; Harrison, 2001; Rich, 2014). By this standard, there are 38 United Nations member states with populations of less than a million (Wettenhall, 2018) (Table 1.2). In the context of small Caribbean islands, microstates are defined by de Albuquerque and McElroy (1992) as those having populations of fewer than 500,000. Thus, demographic size might be geographically relative.

Despite the lack of consensus on how microstates should be defined, Baldacchino

Table 1.2. The world's smallest sovereign states by population, 2020

Country	Capital	Estimated population	Location
Vatican City	n/a	800	Enclaved within Rome, Italy
Nauru	Yaren District	10,800	South Pacific Ocean
Tuvalu	Funafuti	11,342	South Pacific Ocean
Palau	Ngerulmud	21,685	North Pacific Ocean
San Marino	San Marino City	34,232	Enclaved within northern Italy
Monaco	Monaco-Ville	39,000	Mediterranean coast
Liechtenstein	Vaduz	39,137	Alps, between Austria and Switzerland
St Kitts and Nevis	Basseterre	53,821	Caribbean Sea
Dominica	Roseau	74,243	Caribbean Sea
Andorra	Andorra la Vella	77,000	Pyrenees, between France and Spain
Marshall Islands	Majuro	77,917	North Pacific Ocean
Seychelles	Victoria	95,981	Indian Ocean
Antigua and Barbuda	Saint John's	98,179	Caribbean Sea
St Vincent & the Grenadines	Kingstown	101,390	Caribbean Sea
Federated States of Micronesia	Palikir	102,436	North Pacific Ocean
Tonga	Nuku'alofa	106,095	South Pacific Ocean
Kiribati	Tarawa	111,796	North & South Pacific Ocean
Grenada	Saint George's	113,094	Caribbean Sea
St Lucia	Castries	166,487	Caribbean Sea
Samoa	Apia	203,774	South Pacific Ocean
São Tomé and Príncipe	São Tomé	211,122	Gulf of Guinea, off the west coast of Africa
Barbados	Bridgetown	294,560	Caribbean Sea
Vanuatu	Port-Vila	298,333	South Pacific Ocean
Bahamas	Nassau	337,721	North Atlantic Ocean & Caribbean Sea
Iceland	Reykjavik	350,734	North Atlantic Ocean
Maldives	Malé	391,904	Indian Ocean
Belize	Belmopan	399,598	Central America
Malta	Valletta	457,267	Mediterranean Sea
Brunei	Bandar Seri Begawan	464,478	Southeast Asia
Cape Verde	Praia	583,255	Atlantic Ocean, off the west coast of Africa

Source: Central Intelligence Agency (2020); Stato della Città del Vaticano (2020)

and Wivel (2020) emphasize the importance of these miniature countries in the geopolitical landscape of the world. Veenendaal (2013a, p. 440), argues that 'When it comes to...comparative political analysis...the group of microstates is too numerous to be neglected'. Although most published work on small territories and microstates has been done in the context of small islands (Connell, 1988), there are many other non-island contexts that are worthy of investigation.

Dumieński (2014, p. 22) defines microstates as small 'sovereign states that have been able to unilaterally depute certain attributes of sovereignty to larger powers in exchange for benign protection of their political and economic viability against their geographic or demographic constraints'. Likewise, Thorhallsson (2019) suggests that they are small in area and population, have a small administrative capacity, and need some level of external protection. This is the general view of microstates adopted in this book, but as Rich (2014) suggested, physical size arguably plays a more salient role in microstate behavior. This, together with population size are the main criteria used here to define microstates; for the purposes of this analysis, Andorra, Liechtenstein, Monaco, San Marino and Vatican City are included. Three of these are monarchies, which has implications for governance and stability (Veenendaal, 2016b), and Europe's smallest states are relatively affluent compared to small countries in other parts of the world (Wivel, 2020). In all of them, tourism is a major industry and a critical player in their rapid socio-economic development since the Second World War (Bartmann, 2012; Eccardt, 2005; Gelbman and Timothy, 2010). In all of these cases, their status as small, sovereign microstates and the broader implications of their international borders enables them to thrive as tourist destinations (Timothy, 2001c).

While knowingly oversimplifying the notion of state size, this book is concerned with the smallest sovereign states and territories in Europe as defined both by geographical area and population. The discussion on microstates is limited to the five smallest sovereign states—Andorra, Liechtenstein, Monaco, San Marino, and Vatican City. They all have populations of less than 100,000 and physical dimensions of less than 500 km^2. As well, they are not members of the European Union. Although Malta and Luxembourg fit some definitions of microstates, they are not qualified for this book because they have larger populations or physical sizes than the parameters set here, they are full EU members, and they are not dependent upon a patron or caretaker state to the extent the other five countries are. This book is concerned with the most atypical sovereign states and other territories (which also incidentally tend to be comparatively small) in Europe, which have received only scant attention in the tourism literature.

Non-State Polities

Although the discussion so far in this introduction has focused on sovereign states, there are types of territories beyond the normative structure of the state but which are important to understand in political geography and over which certain states exercise sovereign authority. For the purpose of this book, these include autonomous regions, dependencies, overseas territories, exclaves, extraterritorial areas, de facto states, and micronations (Fig. 1.1). Not only is each of these categories legally and politically remarkable, tourism has a significant role to play in how they are ruled and how their smallness or unique geopolitical status is framed and maintained.

Autonomous regions are integral parts of sovereign states that have been granted a high degree of self-rule or self-determination (e.g. Åland and Mount Athos). Certain rules and laws that apply within other parts of the state do not apply in autonomous areas, and the autonomy they possess allows them to legislate many of their own laws and socio-political and economic policies and practices. Dependencies are also autonomous areas, but in the three cases in Europe (Isle of Man, Jersey, and Guernsey), they have an even higher level of independence and are not an integral part of the state, even though they belong to the crown and fall within its general domain. The two overseas territories in Europe—Gibraltar and the British Sovereign Bases in Cyprus—are not an integral part of the UK, but they are its legal sovereign possessions, and the UK legislates on their behalf. Some of Europe's overseas territories and dependences have had opportunities to choose independence and self-sovereignty, but many of them have elected not to do so, largely because the financial incentives for staying outweigh the benefits of national identity. A successful declaration of independence would mean losing aid and subsidies from the metropolitan motherland (Grydehøj, 2018).

Exclaves are small parts of one country that are surrounded by the territory of another country, or are physically connected but can only be accessed by traversing foreign soil. There are a few dozen of these in Eastern and Western Europe, but this book focuses on those that are

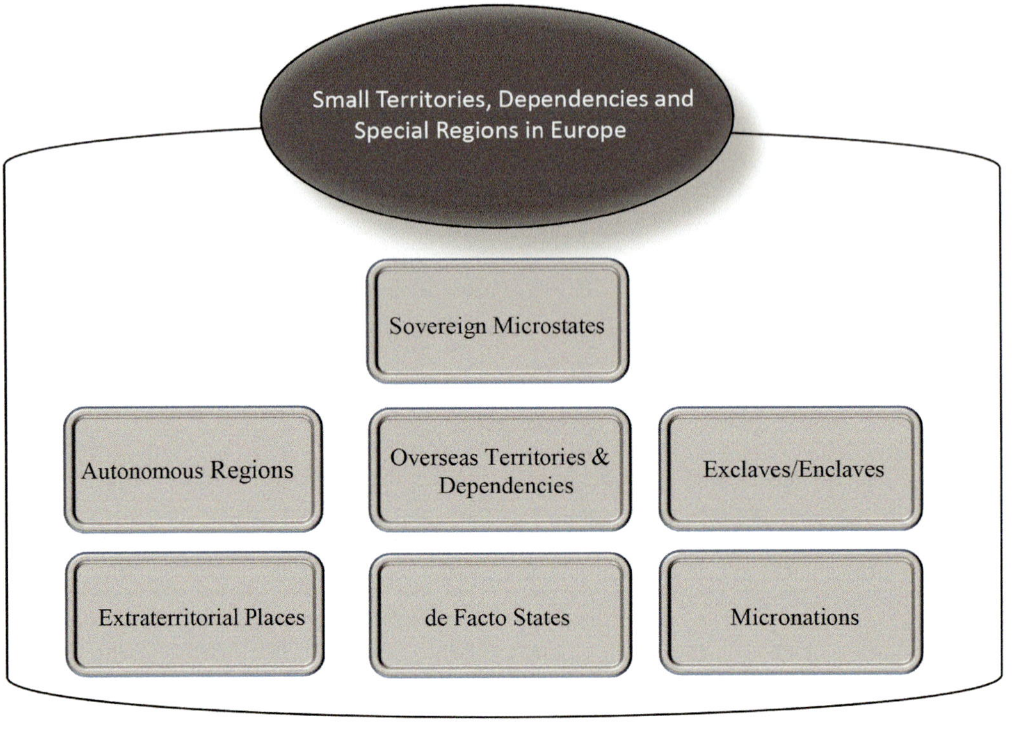

Fig. 1.1. Microstates and Anomalous Political Geographical Polities in Europe

the most tourism-oriented in Western Europe. Extraterritorial zones are fully a part of a state but that state's laws do not pertain to the area in question. This enables foreign entities to operate freely without interference by the host state. Embassies are the best example of this phenomenon but there are others. De facto states, or unrecognized states, are 'countries' that have declared independence from the state to which they legally belong and function as though they were separate countries entirely in most matters of daily life and government administration. The Turkish Republic of Northern Cyprus and Abkhazia are two prominent examples in Europe. Finally, micronations—not microstates—are unofficial and illegal areas that claim to be sovereign entities but have no legal basis for such a claim. European examples include Sealand and Seborga. In most cases, these originate as tongue-in-cheek endeavors, enterprises that aim to satisfy a hobby interest or political maneuver to draw attention to a personal or group cause. They have no legal recognition whatsoever but are a geographical curiosity with tourism interests.

In addition to the current 27 EU member states, there are 22 territories belonging to member states that have special statuses or provisions in their relationships to their European metropoles and to the EU as a whole (Table 1.3). There are three primary categories: Outermost Regions (OMRs), Overseas Countries and Territories (OCTs), and special cases (Kochenov, 2011). OMRs (nine at present) are part of the European Union by virtue of being part of an EU state and through mutual decisions made by the state and the EU. Although part of the Union, they have been granted certain exemptions given their extreme distance from Europe. The six French OMRs (see Table 1.3) are part of the EU, as noted above, but they are outside the Schengen Area and the VAT area. Spain's Canary Islands are party to Schengen, but they are outside the EU's VAT area. Portugal's OMRs, the Azores and Madeira, are part of the VAT area, but they have a lower VAT than mainland Portugal (European Parliament, 2020).

The 13 OCTs are not themselves part of the European Union even though their metropolitan

Table 1.3. European overseas territories/regions

Sovereign state	Region/Territory	National classification	EU classification
France	French Guiana	Overseas Department & Region	Outermost Region
	French Polynesia	Overseas Collectivity	Overseas Country/Territory
	French Southern & Antarctic Lands	Overseas Territory	Overseas Country/Territory
	Guadeloupe	Overseas Department & Region	Outermost Region
	Martinique	Overseas Department & Region	Outermost Region
	Mayotte	Overseas Department & Region	Outermost Region
	New Caledonia	Special Collectivity	Overseas Country/Territory
	Réunion	Overseas Department & Region	Outermost Region
	Saint Barthélemy	Overseas Collectivity	Overseas Country/Territory
	Saint-Pierre et Miquelon	Overseas Collectivity	Overseas Country/Territory
	Saint Martin	Overseas Collectivity	Outermost Region
	Wallis et Futuna	Overseas Collectivity	Overseas Country/Territory
Kingdom of Denmark	Faroe Islands	Autonomous Country	Other/special category
	Greenland	Autonomous Country	Overseas Country/Territory
Kingdom of the Netherlands	Aruba	Constituent Country	Overseas Country/Territory
	Bonaire	Integral part of European Netherlands	Overseas Country/Territory
	Curaçao	Constituent Country	Overseas Country/Territory
	Saba	Integral part of European Netherlands	Overseas Country/Territory
	Sint Eustatius	Integral part of European Netherlands	Overseas Country/Territory
	Sint Maarten	Constituent Country	Overseas Country/Territory
Portugal	Azores	Autonomous Region	Outermost Region
	Madeira	Autonomous Region	Outermost Region
Spain	Canary Islands	Autonomous Community	Outermost Region
	Plazas de Soberanía	Autonomous Cities/other	Other/special category
UK	Akotiri & Dhekelia	British Overseas Territory	n/a
	Anguilla	British Overseas Territory	n/a
	Bermuda	British Overseas Territory	n/a
	British Indian Ocean Territory	British Overseas Territory	
	British Virgin Islands	British Overseas Territory	n/a
	Cayman Islands	British Overseas Territory	n/a

Continued

Table 1.3. Continued

Sovereign state	Region/Territory	National classification	EU classification
	Falkland Islands	British Overseas Territory	n/a
	Gibraltar	British Overseas Territory	n/a
	Guernsey	Crown Dependency	n/a
	Isle of Man	Crown Dependency	n/a
	Jersey	Crown Dependency	n/a
	Montserrat	British Overseas Territory	n/a
	Pitcairn	British Overseas Territory	n/a
	South Georgia & South Sandwich Islands	British Overseas Territory	n/a
	St Helena, Ascension & Tristan de Cunha	British Overseas Territory	n/a
	Turks & Caicos	British Overseas Territory	n/a

states are. They are not directly subject to EU law, but they benefit from their 'associate' status as directed in the 2007 Lisbon Treaty (Murray, 2012). Benefits include the following: their nationals are EU citizens, they are eligible to receive EU development funding, and they cooperate with the EU in many areas of concern, such as climate change, sustainable energy, science and education, and disaster recovery (European Commission, 2020a).

Prior to the UK's January 2020 Brexit, its overseas territories were also eligible for these benefits. There is a great deal of concern in the British overseas territories over their lack of a voice in the 2016 Brexit vote and that the 2020 departure means they will lose funding and support from the EU, as well as EU citizenship. Few have confidence that the UK will take action to support its territories at the same level the EU did. Brexit and the unknowns it inspires have even spurred some independence movements inside the British Caribbean (Staudenmaier, 2018).

There are other special cases that are not OMRs or OCTs but are colonies/territories or exclaves of EU states, and EU laws and policies are applied on an ad hoc basis (Murray, 2012). These include the Faroe Islands, which belong to the Kingdom of Denmark and will be discussed in a later chapter, and the Spanish exclaves along the northern coast of Morocco (Robinson, 1958). These areas, known in Spanish judicial speak as Plazas de Soberanía, include the substantial cities of Ceuta and Melilla, as well as several small coastal islands and a tiny peninsula (Gold, 2000).

Supranationalism

When sovereign states unify for a common economic and/or political cause, in the process giving up some of their sovereignty or autonomy to a central governing body, this is known as supranationalism. There are dozens of supranational alliances around the world, but of most concern in this book are the alliances in Europe, primarily the European Union. This will be addressed in greater depth in the next chapter, but it merits mentioning here in the context of the state, borders, special territories and tourism.

The original premise behind supranationalism is to grow member countries' economies by establishing a single market with a common external trade boundary and to facilitate free trade within the alliance. This would entail the free flow of goods, services, money and people in a way that no longer has to deal with tariffs and duties, quotas, other import regulations, or passports and visas between member states.

In addition to economic growth and fiscal standardization, many supranational alliances today also deal with freedom of human mobility for travel, work and residence, as well as defense and security. In addition to trade in goods and services, the Council of Europe and the European Union are concerned with protecting human rights, democracy, environmental sustainability, education and health care. Supranationalism will be covered in greater depth in the next chapter, but in relation to borders and the anomalous territories that lie at the center of this work, it should be noted here that such alliances and transfer of some elements of state autonomy have significant implications for the growth and development of tourism (Timothy, 2018b, 2019).

Europe's supranationalist goals are ultimately to eliminate the majority of state border functions as described above. The goals also include growing a collective 'super-state' that can more effectively address economic, social, political, and environmental issues at a much larger scale than each country acting on its own, and create a pan-European identity that extends far beyond individual national loyalty and cultural affiliation (Muñoz, 2017; Troitiño *et al.*, 2018). The European Union utilizes tourism as a tool to achieve many of its economic, social and cultural goals.

Tourism

Tourism is a worldwide phenomenon and perhaps the most globalized economic sector of all. It involves people traveling away from home and includes the attractions, activities and services (e.g. transportation, lodging, culinary establishments) that meet the needs of travelers in their home areas before and after the journey, in transit to and from the destination, and while at the destination. Traditional views of tourism tended to lean towards holiday-making or pleasure and leisure, but as a maturing field of study and an evolving system of services, tourism is rightfully being re-conceptualized to be more inclusive of other human mobilities (Hall, 2015; Lew *et al.*, 2015; Timothy and Michalkó, 2016). Scholars now realize that not all tourism is founded upon pleasure-seeking behaviors and leisure-oriented goals. For instance, now accepted into the fold of tourism are pilgrimage, international medical mobility, solidarity visits and volunteering, business and educational travel, shopping and commerce, and dark heritage experiences, among other non-traditional tourisms. People travel for a wide range of motivations and purposes, not just relaxation and amusement. Thus, types of tourism, or niche tourisms, have developed through time and are identified from demand and supply perspectives based upon travelers' motives, expectations and desires, as well as the places they visit and the activities they undertake. Prominent contemporary types include, among others, heritage tourism, sport tourism, coastal and beach tourism, agritourism, ecotourism and adventure tourism.

Since the mid-1900s, international and domestic travel has grown quickly and relatively consistently, with the effects of most political crises, economic downturns, natural disasters, and setbacks not having long-term and permanent impacts on the growth of tourism. However, it remains to be seen how the current (2020) COVID-19 pandemic will continue to affect international travel after it subsides. Travel has become a status symbol and desirable pastime for millions of people throughout the world. Increased affluence and mobility freedom, especially in large markets such as China and Russia, better transportation networks, and increased information and communications technology have brought the world much closer to the traveling public than ever before. The World Tourism Organization (2020c) estimates that 1.5 billion international trips were taken in 2019, the highest number ever recorded.

Although unmitigated mass tourism, which tends to epitomize customary leisure- and pleasure-oriented journeying, has been named the main perpetrator of tourism's negative social, cultural, environmental and economic impacts, tourism of every sort usually results in both positive and negative consequences. Regardless of type and impact, places throughout the world see tourism in all its forms as a significant tool for economic and social development. As such, almost every country, province, county, city, village and rural district wants to cultivate tourism for its economic and social potential.

The small and anomalous entities described earlier are no exception. In fact, in most cases, it is their smallness, quaint character, special status, or geopolitically charged environment that underlies much of their tourism appeal. Most of them gladly capitalize on these distinctions and their inherent competitive advantages for tourism purposes. Even the most contested de facto states and divided territories have an interest in tourism as an avenue for employment and economic growth (Butler and Mao, 1996; Sideri, 2012; Toal and O'Loughlin, 2013). Tourism is either the most important industrial sector in Europe's smallest microstates or among the top few. As well, tourism provides the primary or secondary economic mainstay of most of the continent's exclaves, autonomous regions, overseas territories, and even in several de facto states and extraterritorial zones.

A great deal of tourism research has been undertaken in Europe over the past five decades. However, this work has focused on tourism and hospitality issues and implications almost exclusively in the context of the larger and better-known European countries. This is understandable given that there is more data available, a richer academic literature to draw from, more funding sources, more political latitude for carrying out empirical studies, and generally a wider audience to consume the research. Whereas the larger and more popular countries have received most attention, it is the smallest microstates and other small or unusual polities that rely on tourism the most and present the most unique case studies of political geography, particularly as it manifests in tourism.

Although historically largely overlooked in tourism studies, the crossover between tourism and geopolitics has started to receive increased attention in the literature (e.g. Hall, 2017; Hall, 2017a; Hannam, 2013; Mostafanezhad and Norum, 2016; Timothy, 2001c, 2017f, 2018a, 2020; Timothy *et al.*, 2014). Such a focus is particularly relevant and timely, owing to the extremely contested and dissonant nature of tourism, the ability of tourism to empower or disempower certain populations, the salient power relations involved in tourism development and decision making, and the fact that space and place are highly contested concepts in Europe and elsewhere.

Geopolitics is the study of how geography influences international affairs and power relations between polities, places and peoples. This book is concerned with certain elements of political geography—the area of geopolitics that deals with the state, territory, borders, sovereignty, conflict, and peace. As noted at the outset of this chapter, tourism is all about place. It is the character of place that entices people to visit, and it is the fiber of place that is so often altered through tourism. Geography is the study of place and the fundamental essence of tourism (Timothy, 2018a). This book focuses on political geographical perspectives on place: perspectives that make certain European localities unique and worthy of additional analysis.

Research Challenges

Several theoretical and practical challenges exist in researching microstates, small polities, and anomalous territories. First among these is a general lack of consensus about what constitutes a microstate, an enclave, a de facto state, an overseas territory, an autonomous region and an extraterritorial possession (Caspersen, 2012; Dumieński, 2014; Souleimanov, 2020; Vinokurov, 2007). The multitudes of territorial types, histories, geopolitical relationships, recognition or lack thereof by the community of states, and exceptions to every standard, raise conceptual barriers to understanding sovereignty, autonomy, and power relations more holistically and uniformly. In the words of Dumieński (2014, p. 4), 'the current definitions of microstates are hampered by serious problems of inconsistency, arbitrariness, vagueness and inability to meaningfully isolate *qualitatively* distinct political units'. Thus, each type of polity and territory must be studied individually to understand the power of the state or the powerlessness of the non-state.

Secondly, there is a major data shortage for studying tourism and other social and economic indices in small and irregular territories. This challenge is typically place specific, as some microstates, for instance, collect copious amounts of tourism and economic data, whereas others have almost nothing. Even the available data is sometimes of questionable utility. Where open borders dominate, as in the case of Europe's microstates and exclaves (and among Schengen states), tourist numbers are usually enumerated based on hotel stays. Yet, many entities do not even keep track of these numbers. Thus, hotel data precludes many market segments, including those who stay with friends and relatives, overnight in Airbnb or utilize CouchSurfing services, as well as day visitors. Although the World Tourism Organization (UNWTO) only counts overnight stays as 'tourists' in the strictest sense, tourism in most microstates, exclaves and other unusual polities is defined overwhelmingly by same-day visits. Although large countries have similar data deficits, this situation is especially impactful in small territories where day-trips and transit stopovers comprise the largest components of tourism. Some mini states do their best to estimate day-trippers. Others count only cruise passengers (Monaco), while some (e.g. Liechtenstein) do not even attempt to enumerate day visitors, even though it is their largest tourism market. Thus, the inconsistencies between data collection methods and the data itself create barriers to understanding tourism's impact and growth or decline, or undertaking meaningful comparisons.

The UNWTO counts outbound tourism (people leaving their own country to travel abroad) by arrivals in other countries. These figures are collected by individual countries, then reported to the UNWTO, which enumerates them as outbound travel. As inconsistent as that system is, it is even more problematic from the perspective of microstates. Often, microstate citizens are incorrectly logged as citizens of their larger neighbors. For instance, Canada counts Sammarinese and Vatican citizens as Italians. Other countries fail to distinguish between Monégasque and French citizens or between Liechtensteiners and Swiss.

Data from exclaves is nearly impossible to acquire for the same reasons, and data from de facto states is hard to get because these polities lack recognition and are not included in global tourism totals. Artsakh, South Ossetia and Transnistria, for example, are not recognized by the UN or the UNWTO, and their arrival or stay numbers go unreported, if they are kept at all. Data are often available from dependencies and overseas territories because many of them have their own border controls and are enthusiastic about their tourism sectors.

The third research barrier is, conceptually, certain tourism forms or measures make little sense in the context of microstates and other non-sovereign territories. The dominance of day trips already noted is a case in point. Likewise, domestic tourism is negligible in the European microstates except for recreational activities, which means that domestic tourism, if considered at all, must be conceptualized differently in small states and territories (Canavan, 2012; Timothy, 2020).

Concluding Comments

Many of the geographical phenomena underscored in this book can also be found in parts of Asia, Africa, and the Americas. This book, however, focuses on European cases for several reasons. First, the information about Europe's microstates, exclaves, de facto states, autonomous regions, extraterritorial areas, and territories is plentiful. Second, nearly all of these special places are either dependent on tourism for their fiscal fitness, or tourism is one of the most salient parts of the local economy. Although the focus here is Europe, anomalous territories everywhere can learn from the experiences highlighted in this volume. The island microstates and small territories of the Pacific and Caribbean are particularly drawn to tourism as an economic redeemer, yet they are prone to some of the same vagaries of external forces and geopolitics that occur in Europe. Nevertheless, they also have many concerns and obstacles to development that the European microstates and territories do not have.

The smallest microstates of Europe all depend a great deal on tourism, and so do most of the other anomalous territories described here. The key political geographical concepts outlined in this chapter—sovereignty, borders, states, state sizes, countries, nations, nation-states, secessionism, irredentism, and supranationalism—all play a significant part in understanding the polities examined in the pages that follow. These and many other political geographical concepts have a crucial bearing on Europe's curious polities.

The next chapter deconstructs the notion of transboundary relationships in Europe to provide an understanding of how supranationalism, especially the European Union, influences the everyday lives of residence in Europe's anomalous areas. It also provides a foundation for understanding border issues, economic development and tourism at the national and supranational levels. Chapters 3–7 look specifically at the five mini states (mini microstates) in Europe from a historical, geographical and socio-economics perspective. In particular, these chapters examine the role of tourism in each of these small countries and the opportunities and challenges they face from a largely geopolitical perspective. Chapters 8 through 10 examine some of the manifestations of the political geographical characteristic of other, non-standard state outliers and unrecognized states in Europe and the role that tourism plays in each one, as well as how their geopolitics influences tourism. Chapter 11 consolidates what was learned from the empirical chapters (3–10), emphasizing the problems and prospects of tourism and ordinary life in these unique European polities. Finally, the concluding chapter highlights some of the main concepts elucidated throughout the pages of the book and suggestion future research needs in the area of tourism and political geography.

Europe continues to be one of the most studied and insightful laboratories for understanding geopolitics, globalization, economic development and socio-cultural transformations. Despite a growing interest in the political geography of tourism in Europe (e.g. Hall, 2017b; Singleton, 2016; Zhang, 2019), few scholars have examined the relationships between tourism and anomalous territories, such as mini microstates, exclaves, overseas territories and dependencies, unrecognized states, autonomous regions, extraterritorial places, and micronations. To address this lacuna, this book aims to highlight the plight of the smallest sovereign states and other 'non-standard' territories in Europe in relation to tourism. All of them are heavily involved in tourism, even if they would prefer not to be. Their unique spaces and geopolitical relationships make them extraordinary cases for examining the political geography of tourism in one of the most dynamic regions on earth.

2

Transfrontier Relationships in Europe

Introduction

The previous chapter introduced the idea of cross-border cooperation and supranationalism. Europe is the world's most advanced laboratory for understanding the growing trend in transfrontier collaboration and the integration of states through supranational alliances. The Council of Europe includes all but three countries on the continent, while the European Union is currently comprised of 27 member states. There are sundry other multinational alliances that have created a unique geopolitics in the region as well, all of which have important implications for social, political and economic relationships in Europe. Cross-border collaboration and the supranationalization of Europe have extreme implications for tourism, including trade in goods and services, common currencies and standardized tax systems, environmental protection, infrastructure development, transfrontier human mobilities, and citizenship, to name only a few (Więckowski and Cerić, 2016).

The cross-boundary relationships that continue to grow also have important consequences for the smallest microstates, exclaves, autonomous regions, dependences, overseas territories, and unrecognized countries of Europe. The purpose of this chapter is to establish an understanding of the long practice of cross-border relationships in Europe. The information presented here is important in order to understand the relationships between geopolitics, the anomalous entities in Europe, and tourism. It provides the backdrop to understanding inter-state relationships as they pertain to the development of tourism, sovereignty and other important issues in Europe today.

Cross-Border Cooperation and Bilateralism

In the distant past, there was relatively little cross-border cooperation. In fact, the idea of clearly defined sovereign borders is relatively new in human history. Proto-states during ancient and medieval times were far different to what states are today. Although global trade had existed for centuries to satisfy the needs of the rich, landed gentry and nobility, until the 1800s states essentially acted on their own to produce what they consumed. Rarely did countries collaborate on matters of common concern. As nebulous frontier areas evolved into firm boundaries and as the notion of the modern sovereign state took shape, there was an increasing need for joint collaborative efforts to solve common problems. While historically, local cross-border issues were often addressed through informal interactions and at a small scale, the establishment of the modern state meant that more formalized treaties and diplomatic negotiations were required, even regarding mundane concerns.

Collaboration between polities across national borders is becoming more commonplace in many areas such as trade, law enforcement and security, environmental protection, urban and regional planning, and tourism. Cross-border cooperation happens at several different scales, from two states signing an agreement to many states developing supranational alliances for open trade, with many different layers of transfrontier networking in between.

Bilateral efforts between two sovereign states are pervasive throughout the world, but they are especially common in Europe, with the EU now providing legal frameworks and programmatic initiatives to encourage cross-border governance and collaborative development. One example is through its various INTERREG programs, which started in 1989 as a means of allocating development funds from wealthier parts of Europe to less affluent regions to invest in services and infrastructure. Transfrontier Euroregions administer the funds, many of which have targeted cross-border tourism development projects that were proposed jointly by local borderland authorities (Nilsson *et al.*, 2010; Prokkola, 2011; Więckowski, 2005; Wróblewski and Kasperek, 2019).

Even before the European Union and its modern forerunners, the microstates were already in 'customs unions with their big sisters' (Eccardt, 2005, p. 26). Just as many developing islands states in the Pacific and Caribbean continue to be overly dependent upon their colonial metropoles (Connell, 1991), Europe's smallest microstates have maintained their sovereignty and very existence through their close associations with protective patron states. These bilateral associations between the microstates (Chapters 3–7) and the countries that surround them are often referred to as client-patron relationships, indicating a dependency by the client state (the microstate) on the patron state (its larger neighbor and supporter) for many of its services, most of its military protection, and much of its international diplomacy (Eccardt, 2005; Veenendaal, 2016a, 2020; Wivel and Oest, 2010). Eccardt (2005) points out several concrete examples of run-of-the-mill ways in which microstates are overly dependent on their patron states: postal services, telephone services, utility provision, and diplomatic representation. In Eccardt's (2005, p. 27) words:

> Diplomacy can be a costly undertaking for a microstate—just staffing embassies in almost two hundred countries could make a serious dent in a population, not to mention the costs. So the diplomatic staff of the microstates is limited, and they often rely on their big sisters for consular services...Liechtensteiners around the world can seek help in Swiss embassies, and Sammarinese can go to Italian embassies. Andorra often lets Spain or France represent it at international conferences.

The links between Monaco and France have been described as overly dependent to the point where Monaco, although a sovereign state, has delegated so much of its autonomy to France that it has lost credibility and an equal footing among other states—a common problem among Europe's tiniest countries (Bartmann, 2012; Eccardt, 2005; Forster and Mallin, 2014). Although, Bartmann (2002) hints that there is in the international community a certain sanctity for the sovereignty of even the smallest states. This might be slightly evident with the increasing level of autonomous decision making recently demonstrated by Liechtenstein and Andorra, perhaps suggesting a glimmer of change in the microstate client-patron relationships. In the de facto states described in Chapter 10, the patron-client relationship remains extremely strong (e.g. Russia vis-à-vis South Ossetia and Transnistria) often being the only countries to acknowledge their legitimacy and to support them in the international arena (Kolstø, 2020; Kosienkowski, 2020).

Beyond the transboundary work between local authorities or neighboring states, various multinational alliances have developed in Europe since the early 1900s as a means of dealing with common problems and concerns. This is usually referred to as supranationalism, which is the focus of the rest of this chapter.

Supranationalism

Supranationalism is the condition of member states of an alliance or union devolving some of their authority to a centralized governing body. In joining a supranational union, a sovereign state must transfer some of its sovereignty or internal autonomy to the central organization to which it belongs. The central body then manages certain regional affairs and handles global relations as a collectivity of sovereign states (Timothy, 2019). The idea behind supranationalism is 'strength in numbers'. The common belief is that countries working together in multinational alliances can collectively solve problems, enhance social development, grow national and regional economies, augment human mobility, and ensure the protection of human rights. The primary goals of most supranational alliances today and throughout the twentieth century is to bring down the obstruction effects of international borders, thereby eliminating most trade barriers and customs requirements, effectively turning the blocs into single markets, free trade areas or customs unions—the most common supranational coalitions in Europe.

Customs unions are trade alliances that form a single customs entity. This means that all trade barriers between member countries are eliminated and a common external tariff is imposed on all products entering the union. Thus, once a product enters the union at any location, it can be transferred to any other member state without quotas, taxes or tariffs (Bloom, 2017; Riezman, 2013). A customs union is the most significant underlying rationale for the European Union and the vanguard to the complex coalition the EU has become today. The EU is a customs union, and it is in separate customs unions with Turkey, Monaco, Andorra and San Marino (and currently with the United Kingdom since January 2020), but Turkey is not part of the single market owing to restrictions on the free movement of people between Turkey and the EU, as well as the fact that the union agreement does not include agriculture and certain services (Togan, 2015). Unlike a customs union, *free trade areas* do not impose a common external customs regime, but they guarantee that no taxes, quotas or tariffs are levied on goods and services from one member country entering another (Bloom, 2017; Corbet and Robertson, 2016).

A *single market* is sometimes also referred to as a common market, an internal market or a unified market. It is a multination trade bloc that removes trade barriers (e.g. tariffs, quotas and taxes) on goods and services, develops a unified policy regarding product standards and regulations, and ensures freedom of movement for goods, services, capital and labor (Egan and Guimarães, 2017). In the context of Europe, this is the mechanism that allows EU citizens to live, work and attend school in any other EU state (Bloom, 2017). Because of their small sizes, the microstates have been concerned about their inability to control immigration, which is one of the reasons they have received exemptions from the single market requirement of unfettered abode. Most of the old bilateral mini state-to-protector state treaties already included provisions about free movement across their common borders (Forster and Mallin, 2014).

Although regional alliances between kingdoms were known to have existed in the Middle Ages with regard to such things as fishing rights and peace treaties, the modern-day concept of supranationalism began more recently. The 1834 customs union (*Zollverein*) between the German states was perhaps the earliest example in modern history of sovereign countries collectively forming a full economic union (Keller and Shiue, 2014). Another early example is the development of the Common Travel Area, which ensured unfettered travel between the UK and the newly independent Ireland after 1922. Through several later treaties, the agreement came to include the right of abode of British in Ireland and Irish in the UK, visa-free employment, the right to vote in the other country, free education in both states, and reciprocal social benefits (Butler and Barrett, 2018).

Another early agreement was the Benelux customs union between Belgium, the Netherlands and Luxembourg in 1944 and later the Benelux Economic Union in 1958. Benelux's main purpose was to eliminate trade and customs barriers between the three small countries that had been devastated by the Second World War and ensure the free movement of goods, services and people. By working together as one trading bloc, they believed they could help each other recover from the devastation of the war (Zonneveld and Faludi, 1997)

Council of Europe

As the above examples illustrate, supranationalism in Europe occurs on several different scales, with the largest being the Council of Europe.

Every state in Europe is a member of the Council except Belarus (because it lacks democratic principles and practices the death penalty), Vatican City (because it is a theocracy rather than a democracy) and Kosovo (because it is unrecognized by half the world's sovereign states and there are still concerns about its human rights record). The foremost concern of the Council of Europe (COE) is to uphold human rights, the rule of law, and principles of democracy (Council of Europe, 2020). It accomplishes these goals through a wide range of activities, treaties and conventions, and support to individual states. Although most of the COE's efforts are undertaken at a supranational policy level related directly to human rights and democracy, through its democracy domain, it has established certain activities and policies, which on the ground, aim to enable 'sustainable democratic societies' through protocols and conventions closely related to tourism.

These include the European Landscape Convention, which endeavors to protect exceptional and ordinary landscapes, and encourage cross-border cooperation in landscape conservation. The Berne Convention aspires to protect Europe's wild plants and animals through individual national legislation and transfrontier coordination in areas of research, education, planning and development. The Cultural Routes program attempts to promote cultural diversity and develop a pan-European heritage identity by connecting thematic places in different countries into cultural tourism routes (Timothy and Boyd, 2015; Timothy and Saarinen, 2013). Currently there 38 designated COE cultural routes with almost every country in Europe being represented (Council of Europe, 2020). The mini microstates highlighted in the next four chapters are all members of the COE, while Vatican City is an observer state only. Ironically, San Marino, Liechtenstein, Monaco and the Vatican were not included in any of the cultural routes, but they do enjoy the benefits of COE membership, such as fostering cooperation in areas of culture, nature, democracy, and human rights.

European Union

The most visible supranational body on the continent is the European Union (EU). Taking lessons from Benelux, the European Coal and Steel Community (ECSC) was established in 1951 with membership comprising the three Benelux states plus West Germany, Italy, and France. The aim of this alliance was to create a common market for iron, coal and steel as a means of controlling the supply of materials needed for war to ensure that such an event as the Second World War would not happen again on the continent. The ECSC eventually melded into the European Economic Community in 1957 and later the European Union in 1992 (Bickerton, 2012). As the alliance evolved, so did its aims and purposes. Today, the EU is a customs union, a common market and free trade area, but it also has gone beyond its original economic function to include areas of environmental, social and military concern.

As a customs union, the 27 EU member countries form a single territory for customs purposes. This means that every country applies a collective customs tariff for items imported from outside the EU. No customs duties are levied on products moving between member states, and once goods have been legally imported they can circulate throughout the EU without additional customs checks. This is one of the premier advantages of membership in the EU. It essentially creates both a larger 'domestic' market for individual states' industrial or service output and provides a broader resource hinterland for non-taxable imports.

This and the freedom of movement throughout most of Europe that Sammarinese, Monégasque, Andorran, Liechtensteiner or Vatican passports enable, are the main incentives for the microstates seeking association (or membership) with the EU and its related coalitions (Briguglio, 2016a). One of Malta's main motives for joining the EU as a microstate was to attract additional investments from outside the EU by corporations who might want to use the island state as a 'platform from which to penetrate the European Market' (Briguglio, 2016b, p. 193). Even though Monaco, San Marino, and Andorra have only been 'partially' incorporated into the EU single market through the EU Customs Union (Forster and Mallin, 2014), the same advantage is now available to them through their EEA or EU Customs Union affiliations with the common market (see discussion below). The EU is presently unwilling to consider full membership for

Europe's smallest microstates, because of their limited capacities (geographical and legal capacities) to be able to enact all EU policies and laws regarding the four freedoms within their national borders (freedom of movement for goods, services, capital and labor) (Forster and Mallin, 2014). This especially includes the freedom of abode anywhere within the EU and the freedom to establish businesses and services. As a result, the microstates have opted not to pursue membership but rather to become associated with the common market through other means.

Liechtenstein is seen as a trailblazer among Europe's mini microstates in exercising its own sovereignty beyond the control of Switzerland in recent years, including opening its own embassy in Washington, DC, in 2002 and joining the European Economy Area (EEA) in 1995 to access the EU common market, while Switzerland stayed out of the EEA (Forster and Mallin, 2014; Wivel and Oest, 2010). Andorra has observed Liechtenstein's decisions and may be in a position soon to join the European Free Trade Area (EFTA) and enter the EEA. Because of the impracticality of EU membership, Monaco, San Marino and Andorra currently are in affiliation agreements with the EU single market. Thus, the microstates have the benefits of being party to the single market but are not obligated to satisfy all the requirements of full membership.

Beyond the domain of trade, the EU has come to address other social, economic, ecological and political issues. Citizenship and human mobility, as noted previously, now falls within the domain of the EU. Citizens of EU member states are entitled to two citizenships: their home countries and the EU. Union citizenship entitles people to live, work and attend university in any other EU state. Each country issues a uniform EU passport, which allows its holder to have the same protections and benefits when living and traveling within the single market as any other citizen (Timothy, 2019). Most non-European countries accept EU passports equally, although a few countries (e.g. the United States) still require visas (as of 2020) for Romanian, Bulgarian and Croatian EU passport holders.

Other tourism-related issues of common concern with the EU include environmental protection, heritage conservation, protecting transfrontier cultural and natural resources, a common currency (see below), and the development of regional infrastructure.

Monetary agreements and taxation

The European Economic and Monetary Union (EMU) is an agreement under the general auspices of the EU, which began in 1991 when the European Council decided to enact a set of policies that would integrate the European economies. EMU was initiated in 1992 and includes coordinating state fiscal and economic policies, limiting government debt and deficits, allowing the European Central Bank to enact a unified monetary policy, ensuring that financial institutions in the euro area abide by EU financial regulations, and initiating a common currency, the euro. The EMU's goal is to effect greater economic efficiency, strengthen the EU economy, and provide the benefits of greater scale in terms of a single market in global trade (European Commission, 2020b; Timothy, 2018b).

Although all EU member states participate in the EMU, Denmark opted out of the common currency requirement when it joined the Union and is not obligated to adopt the euro (Naurin and Lindahl, 2010). Several other EU countries do not presently use the euro as their currency. As of 2020, these include Bulgaria, Croatia, Czech Republic, Hungary, Poland, Romania and Sweden. Bulgaria and Croatia recently applied to join the Eurozone, and all of them (except Denmark) are obligated to adopt the euro at some time in the future as per their EU accession agreements.

With regard to the non-EU microstates, the adoption of the euro initially raised significant concerns, foremost because each one utilized the currency of its patron state. Monaco used the French franc, although prior to 1995 it minted a few of its own franc coins. San Marino and Vatican City (each issuing their own lira coins) used the Italian lira. Andorra used both the Spanish peseta and the French franc. Liechtenstein used the Swiss franc, so it was not affected directly by the EU's adoption of the euro. Leading up to the time the euro replaced the currencies of France, Spain and Italy (January 2002), there was considerable concern among the microstates since the EMU was open only to

EU members. In response to this conundrum, in the years leading up to the 2002 release of the euro, bilateral agreements had to be signed between each microstate and the EU so that the euro could become their official currencies. Monaco, San Marino and Vatican City began minting their own euro coins in 2002. Andorra was finally allowed the same opportunity in 2013 and started minting in 2014 (Kalicka-Mikołajczyk, 2018).

In certain exclaves, the adoption of the euro by both the host state and the home state made life more efficient. In Baarle Hertog, where the currency was officially the Belgian franc and in Baarle Nassau, where the currency was the Dutch guilder, for example, commerce became simplified as shops, accountants, and administrative departments in both municipalities began using the same monetary unit. The situation was similar in Llívia. Although the initial adoption of the euro eroded some of the price advantages for consumers in these exclave communities, prices eventually stabilized and as VAT rates and certain other taxes steadied, certain price advantages resumed and have rendered shopping an important activity once again.

All EU members are obligated to levy a value-added tax (VAT) on goods and services. Each country must collect taxes that align with the EU value-added tax code, and each state's contribution to the EU budget derives from the VATs collected domestically (Schratzenstaller and Krenek, 2019). VAT rates vary from country to country, with Luxembourg's being the lowest (17%) and the highest being Hungary's (27%). The EFTA members of the single market are not obligated to collect EU-regulated VATs, but they all levy their own VAT taxes (Norway 25%; Iceland 24%; Switzerland 7.7%; Liechtenstein 7.7%). Being part of the Swiss economic area, Liechtenstein imposes the same rate as Switzerland.

Although not an EU member, Monaco is included in France's VAT regime, where the rate is 20%. Likewise, the British Sovereign Base Areas in Cyprus are not part of the EU but abide by the EU tax policy and are included in Cyprus' tax regime at 19% (Kentas, 2018). Some areas within the EU have been granted VAT-free status as a means of ensuring their economic development. These include the autonomous or semiautonomous areas of Åland, Heligoland, Mount Athos and the Faroe Islands, as well as certain exclaves and pene-exclaves, including Livigno, Büsingen am Hochrhein, Jungholz, and Kleinwalsertal (Cipollini, 2019).

The advantageous relationships that the mini-states and some of the territories have developed with the EU without being members have allowed them to maintain lower VAT taxes or remain VAT-free entirely. This has led to many of them developing into thriving offshore financial centers and safe havens (Paldam, 2013), although the EU has continued to pressure them to ensure their 'offshore' activities are compliant with union laws. Liechtenstein, for example, has long been a known tax haven because of its relaxed banking rules and low taxes, but it was blacklisted by the EU in 2015, until it conceded to more financial transparency.

The British Crown Dependencies of Jersey, Guernsey, and the Isle of Man are well known tax havens. Jersey, for example, does not have inheritance and wealth taxes, which draws many affluent British citizens and their bank accounts (Dharmapala, 2008), although most havens do not require investors and business owners to live physically in the haven or have their businesses set up there (Singh, 2019). Liechtenstein, Andorra, Monaco, the Isle of Man, and the Channel Islands famously host many company headquarters owing to their lower corporate taxes, far more than their own populations could support. There are current global efforts to change tax laws so that taxes are levied where the work is done rather than where a company keeps its headquarters.

The Schengen Agreement

According to a growing pan-European vision of open borders on the continent, a treaty was signed in Schengen, Luxembourg, in 1985 by five states (Netherlands, Belgium, Luxembourg, France and Germany) to enable barrier-free travel between them. In 1990, changes to the agreement established a common visa policy in which any signatory state can issue a Schengen visa that is valid for travel to all participating 'Schengenland' countries. It also established mechanisms for greater cooperation in security and immigration. Beginning on March 26,

1995, all border checkpoints between the original five pioneering states, as well as Portugal and Spain, were abolished. In this sense, the Schengen Area functions much like a single country with no routine passport inspections. Over the next few years, several other Schengen countries joined and abolished their border checks. In 1999, the European Union adopted the Schengen protocol and integrated it into EU law.

Today, 26 European states are party to the Schengen Agreement. Although the accord was incorporated into EU law, a handful of non-EU countries have been allowed to join through their association with the EU as members of the European Economic Area or European Free Trade Association. These include Norway, Iceland, Switzerland, and Liechtenstein. Like Denmark opted out of the Eurozone, Ireland and the UK opted out of the Schengen treaty because of their open border enabled by the Common Travel Area, which had already been established between the two countries in the 1920s. With the UK's withdrawal from the EU on January 31, 2020, Ireland remains the only EU state with an opt-out. As of mid-2020, four EU states have not yet enacted the Schengen Agreement—Bulgaria, Croatia, Cyprus and Romania—but they are obliged to enter the fold as soon as they are able.

The Schengen Agreement does allow individual countries to re-establish temporary border checkpoints during unusual times and in extenuating circumstances. This has occurred on several intra-Schengen borders beginning in 2015 with the Middle East and African refugee crisis. Several countries established border checkpoints and built fences on their frontiers as a means of gaining control over the refugee situation. In 2020, with the COVID-19 pandemic, many additional inner-Schengen border checkpoints were re-established and roads, bridges and sidewalks barricaded as individual states decided to seal their borders to prevent the spread of COVID-19. These measures are required by EU law to remain only temporary until the problems are solved.

In 2008, Liechtenstein signed the Schengen accord and was allowed to join in December 2011 after satisfying certain requirements related to its banking and financial sectors. The other mini microstates are not official members of Schengen. However, because of their longstanding border arrangements with their neighbors, and because Monaco only borders France, San Marino only borders Italy, and Vatican City only borders Italy, these three microstates function as de facto members of the Schengen Area with France and Italy. Andorra is not a party to the Schengen Agreement, so passports are required to enter and exit, and although Andorra does not require visitor visas, the Schengen Area might, so travelers departing Andorra are required to possess a multiple-entry Schengen visa.

Although an EU member state, Cyprus is not allowed to sign the Schengen Agreement until its partitioned status and de facto border situation is resolved. The British Sovereign Bases in Cyprus are also technically outside the EU, so their situation as regards Schengen still needs to be determined. The unrecognized states in Eastern Europe are not legally part of any Schengen countries, and the territories and dependencies are not party to Schengen (Murray, 2012). Although metropolitan countries in Europe may be members of Schengen, their overseas territories generally are not. None of France's outermost regions or overseas countries and territories are included in the Schengen area. Likewise, none of the Netherlands' overseas possessions belong. All of these have their own visa regimes and border controls, typically even when arriving from the motherland (Timothy, 2020). Many of the unique territories examined later in the book (e.g. Åland, Heligoland, and Mount Athos) are party to the Schengen Agreement, even if special permits are required to enter (i.e. Mount Athos).

European Free Trade Association

Members of the European Free Trade Association (EFTA) include Iceland, Liechtenstein, Norway, and Switzerland. EFTA is not a customs union or single market; it is a free trade area, wherein there are no import or export quotas, import taxes, or tariffs on goods or services transferred between participating states (Bloom, 2017). EFTA commenced on May 3, 1960 as an alternative trade alliance for the European countries that did not want to join the European Economic

Community (EEC/EU). In the beginning, this included the current four, as well as Austria, Denmark, Portugal, Sweden, and the UK (Corbet and Robertson, 2016). These latter five eventually left EFTA and joined the EU.

The remaining four EFTA members are all party to the EU single market either through the European Economic Area mechanism (Iceland, Norway and Liechtenstein) or affiliated through a series of bilateral agreements (Switzerland). They are also included in the Schengen Area but not the European Union Customs Union (Corbet and Robertson, 2016). Although Switzerland is a full member of Schengen and is affiliated with the EU single market, it is not fully integrated into the single market, and therefore maintains border inspections with all of its single market neighbors, except Liechtenstein.

European Economic Area

The European Economic Area (EEA) was established on New Year's Day 1994 as a legal framework to extend the EU's single market to the non-EU states belonging to the European Free Trade Association: Norway, Iceland, Switzerland, and Liechtenstein (Lavenex, 2011). However, Switzerland did not join the EEA and is therefore not as fully integrated with the common market as the other three EFTA states are, owing to concerns about losing its economic and tax advantages and losing control over immigration. Through a series of bilateral arrangements where it accepted many EU regulations on agriculture, transportation, freedom of movement, security and law enforcement, Switzerland has been able to participate in the EU single market without joining the alliance.

The EEA is comprised of all EU states, plus Norway, Iceland and Liechtenstein. Likewise, after its withdrawal from the EU in 2020, the UK became a transitional member of the EEA, which has enabled it and its European overseas territories and crown dependencies to continue as part of the single market on a temporary basis. Because of its EFTA membership, Liechtenstein gained full access to the EU single market in 1995 by way of the EEA alliance. By joining the single market, Iceland, Norway and Liechtenstein consented to abide by the common market principles of free movement of labor, goods, capital and services. However, given its small physical and demographic size, Liechtenstein was able to secure a special exemption from the provision for the free movement of people. Although Liechtensteiners can freely move about within the EU, the country has retained its right to limit the number of EU citizens who can live and work there (Forster and Mallin, 2014, p. 3).

The EEA allows EFTA members to avoid participating in all requirements of the EU. For example, they do not participate in the union's fishing and agricultural policies, and they are not required to join the European Union Customs Union with Turkey.

Conclusion

This chapter has described the notion of supranationalism, or the gathering of sovereign states into large trade blocs as a means of increasing internal market strength and external market reach. This chapter sets the tone for many of the relationships discussed in the sections that follow. Supranationalism, especially in the form of the European Union, has special repercussions for Europe's microstates. Although they have not been allowed to join the EU for various size and capability reasons, they have been admitted into the single market through various associations with the European Economic Area, as well as via individual bilateral treaties that qualify them to join Europe's broader customs union and the trade benefits which that affords. These concepts are important to understand when examining the crossover between tourism and political geography in the mini microstates, but what about the other polities discussed in the book?

The human mobility, free trade, VAT tax concepts and a common currency have significant bearings for exclaves, some of which had developed thriving tourism economies based upon certain tax advantages, different exchange rates, and even legal permissiveness. Once their home and host states joined the EU, some of these advantages disappeared.

Nonetheless, they have been entrepreneurial in finding ways to re-invent their border tourism advantages. The separation of some autonomous areas and dependencies from the EU continue to provide a foundation for tourism development, largely based upon tax havens and centers for business development. That certain autonomous regions (e.g. Åland and Heligoland) are part of the EU but beyond its tax rules has enabled their tourism to thrive. The chapters that follow consider the positive and negative consequences of being inside or outside the European Union for microstates and other political entities in Europe.

3

The Pyrenean Co-Principality of Andorra

Introduction

Since the beginnings of human civilizations, the Pyrenees Mountains have formed a barrier between the Iberian Peninsula and other parts of Western Europe. They form a natural frontier between France and Spain from the Atlantic coast in the west to the Mediterranean Sea in the east, with their summits ranging from 905 m at the lowest to 3404 m at their highest point. For millennia, these daunting mountains have been home to pastoralists who practiced transhumance, or seasonal movements of livestock from the valleys and villages to the mountains for summer grazing (Lew *et al.*, 2015). The Pyrenees are home to several cultural and linguistic groups, including French, Spanish, Catalan, Basque, Aragonese and Aranese (Vaccaro and Beltran, 2010), some of which developed uniquely because of their topographic isolation—the same topography that has protected a handful of minority languages and cultures on both sides of the Spanish-French border.

During the Roman Empire, the mountains formed a formidable barrier between the provinces of Hispania and Gallia; the remains of Roman roads in some of the mountain passes continue to be a reminder of the difficult and dramatic mountainscapes that characterize the French-Spanish borderlands. Nestled among the peaks and valleys of the Pyrenees, between its imposing and once-threatening powerful neighbors, France and Spain, lies the tiny principality of Andorra, one of the most remarkable countries in Europe, which has succeeded in utilizing its mountain culture, language (Catalan), and topography as resources for socio-economic development, including tourism. This chapter describes the unique characteristics of this montane microstate and how it has utilized its matchless situation to become a success story among the mini microstates of Europe.

Place and Time

Andorra is physically the largest of the five European microstates examined in this book (468 km^2) and has a larger population (ca. 77,000) than the others. Andorra is an entirely montane country, situated in the Pyrenees, bordering France and Spain (Fig. 3.1). The country's least mountainous area is in the south where the eastern and northern branches of the Gran Valira River merge. This has provided a wide enough river valley for the foundation and recent rapid physical development of the capital city, Andorra la Vella. For administrative purposes today, and historically, the country comprises seven parishes: Andorra la Vella, Canillo, Encamp, Escaldes-Engordany, La Massana, Ordino, and Sant Julià de Lòria. With

Fig. 3.1. Andorra's landlocked position between France and Spain

Source: Central Intelligence Agency (2020)

the exception of Andorra la Vella, each parish has a main town and a number of rural villages.

According to traditional history, the people of the valleys of Andorra were awarded considerable autonomy and rights of liberty by King Charlemagne of the Holy Roman Empire for their efforts to keep the Moors from advancing northward beyond the Iberian Peninsula (Waldman and Mason, 2006). Later, the territory came under the political control of the Count of Urgell and eventually into the possession of the Catholic diocese of Urgell in today's Catalonia, Spain. During the eleventh century, the Bishop of Urgell created a political and security alliance with a Catalan aristocrat, whose estate was later inherited by the Count of Foix (today in France) (Klieger, 2013). In the early thirteenth century, a dispute erupted between the Bishop of La Seu d'Urgell and the Count of Foix about lordship over the valleys of Andorra. To appease the belligerents, a truce was signed in 1278 that created a situation of shared sovereignty over Andorra by the Bishop of La Seu d'Urgell and the Count of Foix, the count's rights eventually being assumed by the French head of state: the president of France (Comas d'Argemir, 2002).

Today, Andorra remains a co-principality, with the president of France and the Bishop of La Seu d'Urgell acting as co-princes who, in accordance with the original 1278 treaty, receive their annual compensation in the form of hams, wine, and bread (Comas d'Argemir, 2002). In 1993, a new constitution was adopted in a democratic referendum. This constitutional change formally increased the power of the electorate, limited the authority of the co-princes and created a full parliamentary diarchy with the 28-member General Council and independent judiciary more fully representing the interests of the people.

Slightly less than half of the country's population are Andorrans; the remainder is composed of foreign residents and expatriate workers. Spanish citizens comprise the largest group of resident foreigners (24.7% of the population), followed by Portuguese (12.6%) and French (4.1%). Most of these live in the country on work permits (Govern d'Andorra, 2017). In terms of religion, most of the country's population are Catholics, although there are also small Protestant, Jewish and Muslim cohorts. Although the official language is Catalan, Spanish, Portuguese and French are also widely used because of the country's large migrant workforce.

Neighborly Relations

Andorra is the only microstate examined here to maintain passport and customs inspections at its borders with the EU and vice versa. Andorra signed a customs union agreement with the EU in 1990, which came into effect in 1991 and ensures strong trade relations between the two parties, especially in the area of industrial goods. However, the principality is not part of the EU single market and is not subject to the EU's tax laws or party to the Schengen Agreement. Because it is outside the EU and Schengen Area, Andorra maintains border controls on incoming traffic. For the same reason, France and Spain maintain customs and passport inspections on behalf of the EU. Andorra accepts all passports and Schengen visas (Emerson, 2007), whereas France and Spain check passports and visas because people coming from Andorra are considered new Schengen arrivals, so travelers must therefore have a multiple entry visa.

The EU inspections at the Spanish and French frontiers also ensure that shoppers

Fig. 3.2. The Andorran and Spanish Border Checkpoint near Sant Julià de Lòria, looking into Andorra

have not exceeded their duty-free allowances (Emerson, 2007). Because there are two entrances to Andorra at Pas de la Casa, and to avoid crowded winter conditions in the popular ski resort, Andorra's border inspections take place 3.3 km inside France at a facility shared with French border officials (Timothy *et al.*, 2014). At the main Spanish crossing, Andorra and Spain both maintain separate passport and customs controls on their respective sides of the boundary (Fig. 3.2). Bilateral agreements also enable the principality to use the euro as its national currency, and in 2014, it began minting and circulating its own euro coinage.

Historically, France has acted on behalf of Andorra in many international settings and in terms of diplomacy. However, with the country's new 1993 constitution, which lessened the power and influence of the two co-princes and strengthened the country's democratic processes, the principality has become more independent in its decision-making and international affairs. That same year, it joined the United Nations and several other international organizations as an independent member and began establishing formal diplomatic relations with many countries, although presently it maintains embassies only in Austria, Belgium, France, Spain, Portugal and the USA. The principality hosts only a French and a Spanish embassy, and its diplomatic relations are maintained with various countries through their embassies in Madrid, Spain.

One example of the country's historical overdependence on its larger neighbors is its postal services. Interestingly, Andorra has never had its own postal system. France and Spain administer that public service equally, with each parish having two parallel post offices: one French and one Spanish, and each one issues its own separate Andorran stamps. This system continues today, and the Spanish and French Andorran stamps continue to be important for philatelists worldwide.

Economy

Throughout history, Andorra's economy was based on subsistence agriculture, as well as the commercial production of cattle and tobacco. Throughout the twentieth century, its main agricultural product and primary export was tobacco, and its tobacco products were well known around the world (Comas d'Argemir, 2010). However, during the past 25 years, competition from cheaper growing areas has reduced the competitive value of Andorran tobacco; as a result, its production has declined considerably. In 1985, 538,264 kg of tobacco were produced. This grew to 919,736 kg in 1995, but in 2016 production had waned to 244,084 kg. Agriculture is still a critical part of the national economy, with 232 farming business still in operation as of 2016 (Govern d'Andorra, 2017). Other key economic sectors include banking and finances, furniture and electronics manufacturing, commerce and retail, and tourism. In spite of its diverse economy, tourism is Andorra's main source of revenue, comprising between 50 and 80% of the state's earnings and employment (Timothy, 2017a). In 2018, Andorra's per capita GDP was $42,052 USD, comparable to many other European countries (United Nations, 2020).

Tourism and Transportation in Andorra

As already noted, tourism is Andorra's most important economic sector. Since the 1960s, the state has invested a great deal of effort and funding into creating a unique tourism brand: small yet grand, accessible yet isolated enough, and a paradise for skiing and shopping (Comas d'Argemir, 2002; Cuevas and Berthet, 2012; Klieger, 2013; Reichart, 1988; Rinschede, 1977; Taillefer, 1991; Timothy *et al.*, 2014). These efforts have transformed a handful of sleepy Pyrenees villages and valleys into a thriving tourist destination.

In 2018, 8.3 million foreign travelers visited Andorra (Table 3.1). In common with the other microstates, the majority of the country's guests are same-day visitors, who cross the border and return to France or Spain the same day (Jenner and Smith, 1993; Timothy *et al.*, 2014). Roughly two-thirds of the country's arrivals are day-trippers, who come to shop, hike, ski and sightsee. However, the number of overnight tourists is growing and largely includes skiers and spa holidaymakers. In 2017 (and again in 2018), despite its small size, Andorra hosted over three million overnight guests (Table 3.1), with an average length of stay in 2018 of 2.38 nights. The largest market segment is, predictably, Spain, whose visitors comprised 71% of Andorra's total foreign arrivals in 2018. This was followed by the French (16%), other Europeans (10%) and non-European nationalities (3%) (World Tourism Organization, 2020d).

Despite its small size, Andorra is substantially larger than the other continental microstates and has a solid resource base upon which its tourism sector thrives. The most obvious of these resources is the Pyrenees Mountains. The country's montane environment provides a natural backdrop for a prosperous winter sports sector and a thriving hiking and biking industry in the summer. Andorra has three primary ski areas: Grandvalira, Vallnord and Naturlandia, which together comprise approximately 318 km of marked ski slopes and 3075 hectares of skiable areas (Govern d'Andorra, 2017, p. 73). The country has gained a Europe-wide reputation as

Table 3.1. Inbound tourism to Andorra (millions)

	2002	2004	2006	2008	2010	2012	2014	2016	2018
Overnight tourists	3.389	2.791	2.227	2.059	1.808	2.238	2.363	2.819	3.042
Same-day visitors	n/a	n/a	8.510	8.135	6.743	5.663	5.433	5.206	5.286
Total international arrivals	3.389	2.791	10.737	10.194	8.551	7.901	7.796	8.025	8.328

Source: World Tourism Organization (2008, 2011, 2015, 2020a)

Fig. 3.3. The Grandvalira Ski Area in Pas de la Casa, Andorra

a high-quality ski destination (Cuevas Contreras *et al.*, 2016; Simpson, 2014; Taillefer, 1991) (Fig. 3.3). During the summer months, the same mountainsides are used for hiking, and a nationwide network of montane trails has been well developed, mapped and waymarked.

The second resource is the country's healing thermal hot springs (Reichart, 1988). The Caldea spa, which opened in 1994, is now one of the largest thermal spas in Europe and attracts hundreds of thousands of health enthusiasts each year who come for the area's restorative properties and to enjoy the thermal waters in a unique national context (Andorra Turisme, 2020).

The country's small but impactful international border is the third resource that has spurred its development into a popular destination (Comas d'Argemir, 2002; Eccardt, 2005; Timothy, 2017a; Timothy *et al.*, 2014). The quaintness of being a microstate has long been part of the country's appeal, but today the border is particularly meaningful in terms of retail and shopping. Border policies regarding imports, exports and taxes have created conditions that have made Andorra a 'shopping paradise' (Reichart, 1988, p. 27). Because Andorra remains outside the European Union, it can sell high-end commodities free of duties and tariffs (Emerson, 2007; Tanganelli and Pou, 2012; Timothy, 2017a). As well, because it is not subject to the EU's tax policies, the local VAT tax (4.5%) is Europe's lowest, considerably less than retail taxes in Spain (21%) and France (20%). Common cross-border shopping products include cigarettes and cigars, alcohol, perfumes and certain food items—all of which are substantially cheaper in Andorra. Given these price advantages, the EU strictly enforces customs allowances at the French and Spanish borders (Table 3.2).

These tax and price differentials have created a retail landscape that is more intensely commercialized than in any of the other microstates. Besides Andorra la Vella's many boutique and high-end shops, the two main border towns,

Table 3.2. Restrictions on imports of key products from Andorra into the EU (France and Spain)

Product	Amount or Value
Perfume	75 grams
Cologne	375 milliliters
Industrial products	up to 900 euros worth
Coffee	1 kilogram
Tea	200 grams
Cigarettes	300 pieces
Cigars	150 pieces
Spirits over 22%	1.5 liters
Spirits under 22%	3 liters
Table wine	5 liters
Powdered milk	2.5 kilograms
Fresh milk	6 kilograms
Butter	1 kilogram
Cheese	4 kilograms
Sugar	5 kilograms
Meat	5 kilograms

Source: Govern d'Andorra (2017)

Pas de la Casa and Sant Julià de Lòria, are home to shopping centers and many individual stores. These retail 'capitals' of Andorra are particularly busy on weekends as Spanish consumers flock into Sant Julià de Lòria and Andorra la Vella by personal car and shopping specific motorcoaches, and French consumers descend upon Pas de la Casa in private cars and chartered motorcoaches (Fig. 3.4). The price advantages between the two EU states and Andorra are enough that even paying for a same-day return bus passage (shopping package) is economically advantageous.

The popularity of the shopping phenomenon led the country's tourism officials to initiate the annual Andorra Shopping Festival, which takes place in Andorra la Vella and nearby towns. The festival draws people from all over the world, emphasizes the country's shopping opportunities, and provides festivities, workshops, street art, gastronomic celebrations and many other celebratory activities that highlight the country's retail, arts and cultural sectors.

A culture that remained relatively isolated for centuries is the fourth main asset for the country's tourism industry (Eccardt, 2005). Its built heritage includes Romanesque churches

Fig. 3.4. Chartered coaches, such as these in Pas de la Casa, bring French shoppers on day-trips to Andorra

Fig. 3.5. The Romanesque Church of St Joan de Caselles is one of Andorra's premier heritage attractions

and bridges, living villages, medieval buildings, and some 20 museums and heritage centers (Figs. 3.5–3.7). Andorra's living culture includes its unique Catalan identity, language and celebrations, as well as a unique mountain gastronomy (e.g. dishes from wild game and foraged plants). These elements of cultural heritage and others are actively commemorated throughout the country. Most parishes have annual festivals that celebrate local specialties and traditions related to food, nature, snowfall and shopping.

Three of the five microstates highlighted in this book are home to UNESCO World Heritage Sites (WHSs)—an important designation and global heritage brand, particularly important for very small states. Andorra's sole UNESCO-listed World Heritage Site is Madriu-Perafita-Claror Valley, located in the south-east and covering approximately 9% of the principality's total area. The combination of unique glaciated mountainscapes and remnants of human activity 'offers a microcosmic perspective of the way people have harvested the resources of the high Pyrenees over millennia. Its dramatic glacial landscapes of craggy cliffs and glaciers, with high open pastures and steep wooded valleys...reflects past changes in climate, economic fortune and social systems, as well as the persistence of pastoralism and a strong mountain culture...The site features houses, notably summer settlements, terraced fields, stone tracks and evidence of iron smelting' (UNESCO, 2020, n.p.). Andorra is extremely proud of this UNESCO designation, and the WHS has become one of the main centers for summertime hiking, fishing and camping.

For many years, the mountainous topography of Andorra kept the country in an underdeveloped state, and it remained a low-key tourist destination. Access to, from and within the country was difficult at best, particularly in the winter when mountain passes were snowbound for months at a time. However, during the past 40 years, considerable government

Fig. 3.6. The medieval village of Pal is one of Andorra's most scenic historic villages

Fig. 3.7. The 16th century Casa de Vall was until recently the home of the ancient parliament of Andorra

Fig. 3.8. The Envalira Tunnel was completed in 2002 to enable travelers to avoid the Envalira Pass

efforts to ease accessibility challenges for the local population and to develop its tourism sector, have been successful in creating a modern, accessible state. Tunnels were constructed to bypass mountain passes, roads were improved, and viaducts were built over steep river valleys and ravines (Timothy *et al.*, 2014) (Fig. 3.8). None the less, there are still physiographic limitations on transportation and infrastructure.

There is no airport within the territory of Andorra, but the country does have an airport. Andorra-La Seu d'Urgell Airport is located in Catalonia, Spain, approximately 12 km from the Andorran border. It was built to serve the air travel needs of the city of La Seu d'Urgell and the principality of Andorra (Suau-Sanchez and Voltes-Dota, 2019). In large part, its purpose was to help grow tourism in Andorra by providing a closer air hub than Toulouse, France, or Barcelona, Spain. In 2010, the newly renovated airport opened to smaller, regional aircrafts and received more than 1700 flights during the first year. In 2014, in an agreement between the Catalan regional government and the Principality of Andorra, and with a significant financial investment from Andorra, the airport received its current name and officially became Andorra's airport. In January 2015, the airport started to receive chartered flights from cities in Spain and France, but the traffic was short lived, as tour operators preferred to fly into Toulouse or Barcelona and transfer tourists by coach (Cuevas Contreras *et al.*, 2016). With the rebranding and reorientation of the airport in 2014–2015, two airlines were formed by private investors—Air Andorra and Andorra Airlines. Neither of them was affiliated with the microstate, although their central operations were to be at Andorra-La Seu d'Urgell Airport. Due to a lack of demand, neither one of these airlines was able to commence operations (Suau-Sanchez and Voltes-Dota, 2019), and at the time of writing in 2020 there were no commercial flights operating in or out of Andorra's airport.

While the airport situation remains in limbo, the country does have international air access directly through its three heliports in La Massana, Arinsal and Andorra la Vella. Although there are no regularly scheduled flights into these heliports, private companies offer transit taxi services to Toulouse, Barcelona and Girona, and sightseeing flyovers.

Andorra has a substantial road network with a total length of 269 km (Govern d'Andorra, 2017). Major road construction in the past half century has provided increased access to all towns and most villages. The national bus network, *Cooperativa Interurbana Andorrana*, as its name indicates, provides services on several routes between major towns and cities. Many villages and settlements, however, are not serviced by public transportation, which means that a great deal of local travel is done by personal vehicle. Uber ridesharing is available throughout the capital city and nearby areas.

Access to the principality by road is rather easy. There are technically three main border crossings and two minor ones. The minor ones lead to very isolated communities in Spain, one of which has no outlet to other communities on the Spanish side, making it (Os de Civís) a pene-exclave of Spain. This concept will be discussed later in the book. The most frequently crossed border checkpoint is at the Spanish border just south of Sant Julià de Lòria. The other two border crossing roads (through one customs checkpoint) are located in the town of Pas de la Casa, at the French border.

Many international motorcoaches traverse the Pyrenees daily between Andorra and various locations in north-eastern Spain and south-eastern France. Barcelona and Toulouse, the principality's main international gateway cities are 197 km and 185 km from the country respectively. Regular bus services connect the microstate with these two large cities and other regional cities and towns, such as Girona, Lleida, Puigcerda, and Perpignan (Cuevas Contreras *et al.*, 2016). There are no train stations in Andorra, but there are two stations relatively close in France, both of which are accessible by public buses (Cuevas and Berthet, 2012).

As tourism is Andorra's main industry, the country has a wide range of accommodations, including spas, hotels, inns, bed and breakfasts, Airbnb rentals, and campgrounds. In 2018, Andorra had 239 lodging facilities in the category of hotels and similar establishments with 11,998 rooms and 32,903 bed spaces (World Tourism Organization, 2020a).

Outbound travel from Andorra

Compared to Liechtenstein, Monaco, San Marino and the Vatican City, Andorra is a rather large microstate with more assets to keep its residents close to home. None the less, Andorrans and the country's foreign residents are avid travelers, with the majority of international visits being made to Spain, France and Portugal. Andorrans also travel frequently throughout Europe, but few European countries record Andorran arrivals, with the exceptions of Italy, Poland and Romania (World Tourism Organization, 2020b). This is primarily because, even though Andorra is not a Schengen state, its citizens are treated in large part as though it were, with visa-free access to all Schengen countries for up to a three-month stay. For the most part, within the EU, numbers of Andorran tourist arrivals are tabulated, if they are at all, from accommodations records. In addition to Andorrans' normal visa-free access to the Schengen Area, France, Spain and Portugal allow Andorran citizens to stay longer for residency or work (European Travel Information and Authorization System, 2020).

In 2018, nearly 37,000 Andorran international journeys were undertaken, although only 4799 were recorded as staying overnight abroad (Table 3.3). These figures do not reflect travel

Table 3.3. Outbound travel by Andorrans[a]

	2000	2003	2006	2009	2012	2015	2018
Number of arrivals of Andorran citizens or residents in foreign countries	3457	2525	6140	5648	16,716	24,909	36,746
Number of Andorran citizens or residents staying at least one night abroad	4235	737	1379	662	1065	2568	4799

[a]Does not include Andorrans arriving in Spain, France or Portugal
Source: World Tourism Organization (2020b)

to Spain, France and Portugal, Andorrans' primary destinations. The numbers in Table 3.3 indicate that international day trips, primarily to Spain and France, are as important for outgoing Andorrans as they are for non-Andorrans coming in. Besides its nearest neighbors, the countries reporting the highest number of Andorran visitors were Mexico, Italy, Belgium and the USA. Many of the principality's immigrant residents travel back to their homelands frequently, especially Spain, Portugal and France, to visit family and friends. For the most part, Andorrans travel throughout the world for the same reasons other nationalities do: leisure holidays, pilgrimages, business travel, visiting relatives, and other such purposes.

Tourism issues and concerns

As previously noted, the montane environment of the high Pyrenees is the foundation of life for the Andorran people, in terms of water, energy production, food, agriculture, and tourism. The importance of the skiing industry cannot be overstated for the principality's economic well-being. Yet here, like other mountain regions of the world, the effects of climate change on snowfall levels is extremely disconcerting (Pons-Pons *et al.*, 2012). Tourism specialists have elucidated how snow deficits will affect ski tourism and therefore damage destinations' reputations for 'good snow'. Forward-looking, scenario planning research suggests that destinations which depend on natural snowfall may have to resort to more snow-making, shorten their ski seasons and reduce the number of slopes available for use, not to mention the spin-off effects of lowered values of holiday property rentals, increased unemployment and increased competition from other areas and tourism sectors. Such effects, according to Steiger *et al.* (2019), will depend on the rate of warming and the adaptive measures taken by skiers and destination managers. In microstates, such as Andorra, where a significant portion of its economy is based on winter sports, such prospects are frightening, and critical decisions have to be made regarding how the principality can be more economically and environmentally resilient in the face of climate change.

A second issue facing Andorra is its interest in integrating more closely with the European Union. An agreement of association is one goal both sides are currently considering, but the exact nature of the association has not yet been worked out. However, suggested options have ranged from a bilateral agreement between the EU and Andorra for an 'associated relationship', the entrance of Andorra into the EEA, which would give it a higher level of regional integration, or full EU membership. Full EU membership is presently the least likely option, as the EU and its legislative frameworks are not currently equipped to cope with very small states. Although Andorra operates border checkpoints with its EU neighbors and vice versa, its borders are easy to cross, and all nationalities are welcome to enter. Some future actions that involve closer affiliations or integration with the European Union would probably not make frontier crossings any easier, but they could diminish the country's tourism competitive advantage, as in the case of shopping.

Accessibility continues to be a significant concern for tourism in the principality. Although the industry remains successful with the country's current transportation and access limitations, many potential visitors find Andorra's relative isolation to be somewhat off-putting. The airport plans have not yet come to full fruition, but this is likely a result of costs outweighing the benefits and economy of scale. It remains more cost-effective to fly visitors into Barcelona or Toulouse and transfer them to Andorra by motorcoach. It remains to be seen if transportation and access limitations can be overcome through more collaborative efforts and reduced costs.

Like most popular tourist destinations, overcrowding is a problem in Andorra la Vella during high tourist seasons and during urban festivals (Fig. 3.9). Likewise, in Pas de la Casa, one of the Pyrenees' most significant ski resorts, some 15,000 winter sport and shopping enthusiasts descend on the village each day during ski season, creating severe crowding in town and on the slopes (Timothy *et al.*, 2014). Likewise, the rapid construction of new buildings in Andorra la Vella and Pas de la Casa and their environs for the past 25 years may have future implications for overdevelopment and overcrowding, leading to visitor disappointment, if not managed wisely (Emerson, 2007).

Fig. 3.9. Modern Andorra la Vella can be extremely crowded during ski season and during the city's shopping festival

Conclusion

Andorra is now a major tourism venue. Despite the obstacles the Pyrenees Mountains create for access and infrastructure development in the co-principality, the very same montane topography provides the essential foundation for its tourism success and its unique sense of self and national solidarity. Its unique status as a co-principality and microstate and its subservient relationship with Spain and France have created much of the travel mystique assigned to the mountain state, yet it also means that Andorra did not receive universal recognition as a fully sovereign state until 1993, which enabled it to join the UN and other international bodies and negotiate its own treaties and diplomatic relations. Its special relationship with France and Spain has enabled a thriving cross-border flow of regional tourism and a large-scale foreign workforce to labor in the tourism and hospitality sectors.

It has become a well-regarded ski destination in Europe, a well-publicized shopping center, and a utopian hiking and camping destination. Its status as a mini state enhances the value of all of these products, and its non-membership in the EU continues to provide the advantage needed to maintain a successful offshore banking industry and 'tax-free' commercial zone. Although the country now has access to the European single market through its customs union with the EU, discussions are under way to conceive of ways to become even more closely integrated with the European Union. Such efforts must be carefully considered, for closer relations may reduce Andorra's competitive advantage for tourism.

4

Europe's Alpine Mini State: The Principality of Liechtenstein

Introduction

Liechtenstein is one of the wealthiest countries in the world. It has one of the highest standards of living and has a long history of agriculture, industry and sovereign independence. Tourism has been somewhat of an afterthought for past princes, legislators, and the community at large. Although there has long been a tourism sector, it was traditionally seen as a slender spot in the microstate's economy. A ski resort was developed in Malbun in the 1960s, but it has been difficult to compete with better known and more accessible ski areas nearby in Switzerland or Austria. The resort became popular during the 1970s and 1980s. However, tourism in Liechtenstein traditionally was defined by small numbers of motorcoaches stopping by each day as part of Alpine package tours during the summer months with a few dozen Americans, Canadians or British tourists descending on the tiny capital, Vaduz, for 45 min or an hour to look around and buy a souvenir from a place many of them had never heard of before. The scenery was beautiful, but could it differentiate Liechtenstein from its Swiss, Austrian or German neighbors?

Today, the Principality of Liechtenstein is a thriving tourism destination, and national administrators are keen to target tourism as part of an increasingly diverse and affluent economy. Although the drawback of short-stay visits continues to dominate the tourism scene, the country has become a well-regarded destination with a well-planned and marketed reputation for skiing, culture, nature trekking, mountain climbing and culinary experiences. This chapter focuses on this unique alpine microstate and how tourism manifests there. In particular, the chapter examines how Liechtenstein's transfrontier relations, particularly its special relationship with Switzerland, affect the country's tourism and socio-economic development.

Place and Time

The tiny Principality of Liechtenstein (160 km^2) is nestled between Austria and Switzerland. The country is 25 km at its longest and almost 11 km at its widest (Fig. 4.1). It is comprised of two physiographic regions: the Alps and the Rhine River lowlands, with the majority of the population (39,137) living in the lowland area (Fig. 4.2). The capital and second largest municipality is Vaduz, which is home to 5700 inhabitants and the center of Liechtenstein's tourism activity. Prince Hans-Adam II, the wealthiest monarch in Europe, rules the microstate from

Fig. 4.1. The tiny principality of Liechtenstein is nestled in the alps and Rhine River valley between Austria and Switzerland (Source: Central Intelligence Agency, 2020)

Vaduz Castle, overlooking the capital. For administrative purposes, Liechtenstein is divided into eleven municipalities, most of which include a main town and a rural hinterland: Balzers, Eschen, Gamprin, Mauren, Planken, Ruggell, Schaan, Schellenberg, Triesen, Triesenberg and Vaduz.

The Germanic Alemanni people settled in the area of today's Liechtenstein around the fifth century AD, and by the twelfth century, German had become the established language. During the Holy Roman Empire, the area of today's Liechtenstein comprised the Lordship of Schellenberg (established 9th century) and the County of Vaduz (established 1342), both of which were ruled by minor nobility until they were purchased by Prince Johann Adam Andreas of the House of Liechtenstein in 1699 and 1712 (Beattie, 2012). In 1719, the two provinces unified and together became the Principality of Liechtenstein, named after the royal house that owned them. Under Napoleon's

Fig. 4.2. Most of Liechtenstein's population resides in the Rhine River Valley (left). Switzerland is located on the opposite side of the river

titular rule, Liechtenstein became an independent state and was a constituent monarchy in the Confederation of the Rhine from 1806 to 1815. The German Confederation replaced the Confederation of the Rhine and lasted until its dissolution in 1866 (Moore, 1960; Raton, 1970). Although Liechtenstein was a sovereign state in various forms after 1806, it finally became a fully independent state once again in 1866 after the disbanding of the German Confederation.

Liechtenstein is a constitutional monarchy, wherein political power is shared between the ruling monarch and the democratically elected *Landtag* (parliament) and the cabinet, comprising the prime minister and other executive officers/ministers (Büsser, 2016; Corbett *et al.*, 2017; Łukaszewski, 2015). The prince continues to wield more decision-making power and legal influence than most contemporary European monarchs (Wolf, 2015, 2016a, b). Most Liechtensteiners are Alemannic and speak German, albeit with unique Alemannic or Walser dialects. Approximately 30% of the country's population are immigrants from Switzerland, Austria, Germany, Italy and Turkey. Non-Liechtensteiners comprise 70% of the country's workforce, many of whom commute across the border daily from Austria, Germany and Switzerland (Office of Statistics, 2019).

Roman Catholicism is the official state religion, although there is a significant population of Protestants and a small minority of Muslims. The combination of devotion to the monarchy, the country's small size and independence, the local German dialects and its economic prominence in Europe, which far exceeds its physical size, has created a unique Liechtensteiner identity that is unmatched in many of the larger countries that surround it (Ostergren, 2010).

Neighborly Relations

Between 1852 and 1919, the principality was in a customs union with the Austro-Hungarian Empire, which bordered the country on the east and north. However, with the dissolution of Austria-Hungary in 1918–1919, Liechtenstein joined a customs and defense union with Switzerland, which remains in effect today. The

Fig. 4.3. The open Swiss-Liechtenstein border as marked on a Rhine River bridge

two countries form a single economic and currency union and maintain a completely open border (Fig. 4.3). Liechtenstein uses the Swiss franc, and its economy is closely tied to the Swiss economy. The microstate's postal services, while minting its own stamps, is operated by the Swiss postal service. Liechtenstein maintains direct diplomatic relations with several countries but only has embassies in Switzerland, Austria, Belgium, Germany and the USA. Elsewhere, Liechtenstein's interests are managed by Swiss embassy officials, and Switzerland holds the power to sign treaties on behalf of the principality when invited to do so (Duursma, 2006). Likewise, with Liechtenstein having no military, Switzerland is obliged to protect its microstate neighbor, although Swiss troops are not permitted to enter the country.

The border between the two countries has been completely open since 1924 for the transport of goods, services and people (Moore, 1960). Switzerland has traditionally operated customs and passport inspections at Liechtenstein's border with Austria, and because both countries are now signatories of the Schengen Agreement, along with Austria,

Fig. 4.4. This small Austrian customs office at the Liechtenstein border is occasionally staffed for border inspections

passport and visa inspections have been abolished. Liechtenstein, a long-time member of the European Free Trade Area (EFTA), is part of the European Economic Area, which provides access to, and affiliation with, the EU single market and requires adherence to certain EU laws (Duursma, 2006; Forster and Mallin, 2014; Frommelt, 2016). However, Liechtenstein is not a member of the EU Customs Union, so customs inspections still apply on its boundary with Austria (Fig. 4.4).

Like the postal service, the customs services are provided by Switzerland. Many of Liechtenstein's diplomatic issues are dealt with by Switzerland, and its larger neighbor controls currency and monetary policies. Some judiciary responsibilities are carried out by Switzerland on behalf of its small neighbor, and long prison sentences are enforced in Switzerland. The microstate has delegated many other government responsibilities to Switzerland, which has caused many other countries and the United Nations even to question its status as a sovereign state at different times, until it was admitted in 1990 (Bartmann, 2012; Neumann and Gstöhl, 2006). Although the microstate is a UN member, it does not participate in all UN-affiliated organizations, including UNESCO.

Although Liechtenstein's individual foreign relations have seen some growth beyond its relationship with Switzerland, the principality maintains only a few embassies, in Austria, Belgium, Germany, and Switzerland, as well as in the USA. Because of its small size, not all countries with which Liechtenstein has bilateral relations host a Liechtensteiner embassy. In all other countries, the country's diplomatic affairs are managed by Switzerland's resident ambassadors. Liechtenstein does, however, maintain a handful of honorary consulates abroad in key countries such as the USA (four honorary consulates), India, Singapore and Hong Kong. Liechtenstein, with Vatican City, is one of only two sovereign states not to host foreign embassies or other diplomatic missions. Most countries' representatives in Switzerland, Austria or Germany are authorized to handle negotiations with Liechtenstein. Despite the lack of official diplomatic missions, at present, there are 31 honorary consuls located inside Liechtenstein, but these do not function in a position of official diplomacy.

Economy

Liechtenstein remained an underdeveloped agrarian state for the first half of the twentieth century. Following the Second World War, Prince Franz Joseph II led Liechtenstein into a period of rapid economic growth and industrialization (Beattie, 2012) to the point where today, the country has the second highest per capita GDP in the world ($179,258 USD) after Monaco (United Nations, 2020). The main economic sectors of the principality are services (e.g. banking, technology, finance and tourism), manufacturing (e.g. small machinery, dental products, ceramics, textiles, pharmaceuticals and food products) and agriculture (e.g. grains, livestock, dairy products and potatoes). There are presently 55 dairy farms and a grand total of 102 farms throughout the country (Office of Statistics, 2019).

Tourism and Transportation in Liechtenstein

Tourism plays a lesser economic role in Liechtenstein than it does in the other four mini microstates of Europe, although it is none the less very important. The government intends to elevate the role of tourism into the future as it develops the country's international reputation and brand unique from its neighbors (Fehlmann *et al.*, 2002; Passow *et al.*, 2005; Stringer, 2013).

Table 4.2. Top tourist markets in Liechtenstein, 2018

Country of origin	Number of overnight tourists	Share of total tourist stays
Switzerland	24,176	28.35%
Germany	20,756	24.34%
USA	4132	4.93%
Austria	3826	4.49%
UK	3387	3.97%
Italy	2668	3.13%

Source: (World Tourism Organization, 2020a).

In terms of tourism employment, Liechtenstein only records the number of people working in hotels and other lodging facilities, rather than in transportation and food services, as many other countries do. According to 2018 data, only 300 people were employed in the accommodations sector, which comprised 96 lodging establishments, not including campgrounds (World Tourism Organization, 2020a). Besides traditional accommodations, there are 85 Airbnb rentals as of January 2020.

Liechtenstein does not tabulate same-day international visits, but it does report overnight stays. As Table 4.1 illustrates, there has been remarkable growth in the number of overnight tourists in the country, from 48,700 in 2002 to 85,300 in 2018. Not surprisingly, the principality's main overnight tourist market is the Swiss, with 24,176 in 2018, followed by Germans (20,756), Americans (4132), Austrians (3826), British (3387) and Italians (2668) (Table 4.2).

International same-day visits are a more salient part of tourism in the principality than overnight stays in terms of arrivals and economic impact (Jenner and Smith, 1993; Timothy, 2017c). In common with the other four microstates, day-trippers account for the largest number of international arrivals (Volgger *et al.*, 2015), especially from nearby areas of Germany, such as Konstanz; Switzerland (e.g. Buchs, Sargans, Chur,

Table 4.1. Inbound tourism to Liechtenstein (thousands)

	2002	2004	2006	2008	2010	2012	2014	2016	2018
Overnight tourists	48.7	48.5	54.9	58.5	64.3	62.4	61.3	69.1	85.3

Source: World Tourism Organization (2008, 2011, 2015, 2020a)

Fig. 4.5. One of the main tourist shopping streets in Vaduz, the capital city

Zürich and St Gallen); and Austria (e.g. Feldkirch, Bregenz and even as far away as Innsbruck and Landeck). Despite its importance, there is no way to measure day visitors accurately given the open border with Switzerland and the near-open border with Austria. Many package tours that include Switzerland and Austria also include Liechtenstein as a bonus country to visit, although generally bus tours only drive through without stopping or stay in Vaduz for 1 or 2 hours, where tourists are encouraged to shop, walk around the tiny capital or make their way to the prince's castle for a brief visit (Figs. 4.5 and 4.6).

Liechtenstein has a well-developed road network of around 630 km of roads, streets and highways, and the country has the sixth highest ratio of cars per capita in the world. Most ridesharing companies, including Uber and Lyft, are not permitted to operate in Liechtenstein as they do in neighboring Switzerland and certain cities in Austria. A comprehensive national bus service, Liechtenstein Bus, provides transportation access to all corners of the country in all eleven municipalities (Fig. 4.7). The public buses also grant frequent connections to railway services in Austria and Switzerland (Office of Statistics, 2019). The country does not have its own railroad company, but its four stations are served on certain days of the week by the Austrian Federal Railways on its Feldkirch (Austria) to Buchs (Switzerland) passenger line, which passes through 9.5 km of Liechtenstein in the north. Several long-distance international train services pass through Liechtenstein but do not stop (Eccardt, 2005; Timothy, 2017c).

The nearest international airport is in Zürich, Switzerland, from which there are frequent bus services, rail connections and excellent highways. Buchs, Switzerland, is the nearest major railways transit station with an intricate rail network throughout Switzerland and the rest of Europe. There is a heliport in the village of Balzers, which can accept any helicopter arrivals from other Schengen countries. The local company, Rhein Helikopter, offers

Fig. 4.6. The princely castle of Vaduz overlooks the capital city

passenger services to and from Liechtenstein, as well as sightseeing overflights in the area.

Winter skiing at the principality's only ski resort, Malbun, is an important part of the tourism sector (Volgger *et al.*, 2015). During the summer months, the same mountains offer hiking trails and campsites (Fig. 4.8). Cultural heritage is an important part of the country's tourism assets (Figs. 4.9–4.11), but as a non-member of UNESCO, Liechtenstein does not have any World Heritage Sites. None the less, it does have a proud history and unique heritage that manifests in several museums, art collections and galleries, castles and castle ruins, and historic houses. Although wine has been produced in the country since the early Middle Ages, it has become a national passion since the 1970s, and the country also has a nascent wine tourism sector, which partners well with locally produced cheeses and meat-based culinary specialties (Embassy of Liechtenstein, 2020).

Because it belongs to the EEA, Liechtenstein benefits from low trade barriers with the EU, but by not being part of the EU, its tax rate remains outside EU tax controls. This results in relatively low corporate and VAT tax rates (7.7%) that are in harmony with Switzerland's VAT rates (International Business Publications, 2013, 2017). With its advantageous tax benefits, many companies have located within the state's borders, resulting in a growing business tourism sector as well. In common with the other microstates, the production and sales of postage stamps, which are a major collector's item, is an important part of the tourism sector, together with the country's philatelic museum.

Outbound travel from Liechtenstein

Physical proximity, easy access, open borders, and the extremely close political, economic and

Fig. 4.7. Liechtenstein Bus provides regular transportation throughout the principality

Fig. 4.8. Malbun, Liechtenstein's only ski resort, is also a popular summer destination

Fig. 4.9. Gutenberg Castle in Balzers is one of Liechtenstein's most prominent heritage attractions

Fig. 4.10. Traditional Walser homes are a major heritage asset, which are under strict protective legislation

Fig. 4.11. The 15th century Red House in Vaduz is one of the country's main heritage attractions

social relationships between Liechtenstein and Switzerland, explain why the most visited foreign country is Switzerland, which recorded 16,942 Liechtensteiners staying at least one night in the country in 2018 (Table 4.3). Residents of the principality travel to Switzerland to shop, dine and visit relatives. Likewise, being the main gateway for international air transportation for Liechtensteiners, Zürich is a major destination and transit point for those who hail from the microstate next door. Buchs, Switzerland, serves the same purpose for international train connections.

Austria keeps no records of Liechtensteiner visits, but according to officially reported numbers, Italy is the second most popular destination for citizens of this microstate. Day-trips by Liechtensteiners to Switzerland is essentially unmeasurable. As noted earlier, the microstate has a completely open border with its western neighbor, uses the Swiss franc as its currency, and shares a common language and similar dialects. As a result, cross-border trips are extremely common, for some people a daily occurrence, and although these excursions are international, they are generally perceived to be very local.

Table 4.3. Outbound travel by Liechtensteiners

	2000	2003	2006	2009	2012	2015	2018
Number of arrivals of Liechtenstein citizens or residents in foreign countries	8472	9379	14,698	20,175	17,485	20,975	16,426
Number of Liechtenstein citizens or residents staying at least one night abroad	30,812	13,883	15,427	17,184	14,783	17,210	33,369

Source: World Tourism Organization (2020b)

Tourism issues and concerns

Liechtenstein has relatively few concerns related to tourism. However, one issue is its over-reliance on Switzerland as a market source and also with regard to its economic and monetary policies. Because of its close contractual ties to Switzerland and use of the Swiss franc, it is beholden to its larger neighbor with regard to monetary policies, such as a harmonious VAT strategy and banking oversight. This has implications for tourism. For instance, Liechtenstein is constrained in pursuing a retail tourism course of action, if it desired to do so. With a 7.7% VAT tax that matches Switzerland's tax rate, it cannot compete with the larger Swiss economy for German and Austrian shoppers whose VAT rates are 19 and 20% respectively. As well, because of the strength of the Swiss franc, prices are quite expensive in the principality, just as they are in Switzerland.

Liechtenstein does not suffer from a bad tourism reputation. In fact, it has almost no reputation at all in the global tourism marketplace. Recent efforts to stimulate increased visibility have been a measured success (Passow *et al.*, 2005), but the country is still primarily a 'go-through' country rather than a 'go-to' country, despite the fact that it has many tourism assets and enough charm to merit longer stays. Its historical approach to economic development has focused on industrial development and construction at the expense of tourism, which has caused the country to lag behind its neighbors and other European microstates in the realm of tourism. Most tourist visits are very quick stopovers on guided tours. Few package tour participants see anything beyond the shopping streets of Vaduz, and their stays are so short (1–2 hours) that they have little time to make much of an economic impact or enjoy the cultural heritage and nature of which Liechtensteiners are very proud. Thus, not only has Liechtenstein had a difficult struggle for recognition within the legal and policy framework of the European Union, it has also struggled to find its place in European tourism. The principality's historical lack of tourism emphasis has also resulted in an over-reliance on foreign workers and a local population unprepared, untrained and perhaps unwilling to work in the hospitality and tourism sector—a common problem in other microstates, not only in Liechtenstein (Baldacchino, 1994, 1997).

Finally, because Liechtenstein is one of the wealthiest countries in the world and owing to its belonging to the Swiss franc economic regime, the cost of living is high and land values are decidedly inflated. High land values throughout the country raise many challenges for small and medium enterprises (SMEs) in tourism to flourish and make it extremely difficult for new start-ups to open for business.

Conclusion

Liechtenstein, like the other mini states examined in this book, has an overly dependent relationship with its larger guardian. This was by choice in the early 1900s—a choice the country continues to sustain, mostly for economic reasons. By all accounts, Liechtenstein is a major success story among microstates. Although its close relationship with Switzerland has many advantages, the same relationship cost the microstate many years of recognition in the community of states. Its 1920 rejection by the League of Nations began a slow process of diplomacy that finally saw it being able to ascend to full UN membership in 1990. Since that time, it has asserted more autonomy outside its Swiss relations, which may have significant implications for its economic and political future. Its opening of an embassy in Washington, DC, in 2002 and its joining the EEA without the apron strings of Switzerland attest to its growing maturity as a sovereign state in the global arena.

Close relations with Switzerland are advantageous in many ways, but from a tourism perspective, the advantages are largely threefold. First, Swiss tourism authorities (Switzerland Tourism) does most of the principality's international marketing. The official Switzerland Tourism webpages have entire sections devoted to Liechtenstein and its attractions and all Swiss tourist maps, guidebooks and brochures include segments on Liechtenstein. Despite the obvious marketing benefits, this might contribute to ongoing confusion among potential visitors that

Vaduz is just another small town in Switzerland. Even though its tourism agency, Liechtenstein Marketing, maintains its own webpages and prints brochures, these are not as widely distributed as Swiss guides are. Thus, in promotional terms, Liechtenstein may lack its own tourism identity.

The second tourism advantage relates to infrastructure. Because of its small size, the tourism infrastructure in the country is extremely limited. Although it has an excellent road system and a number of Rhine River bridges that connect it to Switzerland, it has not needed to attempt to build its own airport or international railway station. Its international gateways are located and maintained in Switzerland. Third, the open border between Liechtenstein and Switzerland since 1924 has meant steady streams of Swiss and international tourists on overnight or same-day side-trips from Switzerland have been able to cross the border freely.

In common with San Marino and Monaco, it is difficult to persuade visitors to stay overnight in such a small country, especially when neighboring countries have similar but larger product bases for tourism. Liechtenstein's competitive advantage is not its alpine landscape, its ski resort or its heritage sites. These abound close by in Switzerland, Austria and Germany. The principality's competitive advantage is its small size and its very existence as a sovereign state.

5

Large Luxury and Tiny Grandeur: The Principality of Monaco

Introduction

Tourism in tiny Monaco is disproportionately prodigious relative to its physical size. It is a well-regarded destination and one of the most popular on the Riviera. However, it is not suited for everyone. In fact, the country's tourism policy, whether intentional or not, precludes the types of tourists that most Mediterranean destinations try to attract: sun, sea and sand holidaymakers and mass cruise passengers. Although Monaco does accept cruise ships, they tend to be more high-end specialty cruises, which is the nature of tourism in this small principality. Monaco is best known for its luxury yachting, extravagant shopping, sophisticated casino gaming, Formula 1 racing, and high-end real estate—all of which tend to attract the world's jet set who visit largely because they are expected to. The country's lavish built environment, crowded demographics and excessively expensive real estate create an atmosphere and cost of living that prevents most potential tourists from staying more than a few hours which, given space limitations, may be seen as a benefit.

Monaco has a long and special relationship with France, and it is often treated as part of France among foreign tour promoters. As well, France handles many of Monaco's international affairs on its behalf, including some elements that pertain directly to tourism. This chapter describes the unique relationship between Monaco and France, as well as between Monaco and the rest of Europe and how these pertain to tourism. It examines the microstate's unique geographical and social characteristics and its unique tourism products that derive from its small size, location and geopolitical situation.

Place and Time

Monaco is the world's second smallest sovereign state in area (2 km^2) and the sixth smallest in population (38,800). It is roughly 3.3 km in length, from east to west, and 1.1 km at its widest (north to south). Surrounded by France on three sides (Fig. 5.1), the city-state is situated on a series of rocky outcrops on the Mediterranean Côte d'Azur, between the French cities of Nice and Menton, near the Italian border. Monaco has traditionally been divided into five districts based on their primary functions and characteristics: Monaco-Ville (old city and center of government), La Condamine (port area), Monte Carlo (resort and casino district), Fontvieille (leisure and industrial district) and Moneghetti (residential suburb). Today, for administrative purposes, the microstate

Fig. 5.1. The Principality of Monaco is the only non-landlocked mini microstate in Europe

Source: Central Intelligence Agency (2020)

is divided into nine wards: Fontvieille, La Colle, La Condamine, La Rousse/Saint Roman, Larvotto, Les Révoires, Monaco-Ville, Moneghetti, Monte Carlo and Saint Michel.

Given the scarcity of land and the need to accommodate economic growth and a burgeoning population, Monaco has undertaken considerable land reclamation works since the 1880s. The beach area of Larvotto was reclaimed in the 1960s, while the whole of the Fontvieille ward was created entirely from land fill in the 1970s and 1980s (Anthony, 1994). The main port, Port Hercule, was expanded in the 1960s and early 2000s to accommodate a growing yachting sector (Fig. 5.2). Despite the current prince's concerns about sustainability and the quality of Monaco's coastal environment (d'Hauteserre, 2001; Grimaldi, 2006), land reclamation work is currently underway to extend the country's territory further into the Mediterranean by six hectares to accommodate additional luxury apartments, leisure facilities and retail services (Bernouin, 2020; Davies, 2017).

Fig. 5.2. Port Hercule with its collection of yachts

Monaco is said to be the oldest continuously ruling monarchy in history (King, 2008, p. 37). Until quite recently, Monaco was an absolute monarchy. Today, however, Monaco functions as a modern constitutional monarchy, which places certain limitations on princely powers and empowers the population in an open democracy. The Grimaldi family has ruled Monaco since 1297, and the current monarch, Prince Albert III, is a strong supporter of the state's tourism development efforts.

Monaco has a special relationship with its ever-present neighbor, France. Since the thirteenth century, the main superpowers of Europe have recognized the ruling Grimaldi family's sovereignty over Monégasque lands on the Mediterranean coast. From 1793 to 1814, Monaco was under French control and a Sardinian protectorate from 1815 to 1861. In 1848, the two largest portions of the country (Menton and Roquebrune) seceded, leaving only a small portion of centuries-old Monaco intact (Lyon, 2010). In 1860, France annexed Menton and Roquebrune. Out of fear that Monaco could be next and in an effort to demonstrate acceptance of the loss of most of its territory to the French, Prince Charles III forged an agreement with France in 1861 to guarantee the latter's recognition of Monaco's sovereignty and independence (Grinda, 2006). Subsequent bilateral treaties have ensured the continuation of Monégasque independence and set the tone for the political, social and economic relationships between the two neighbors.

Home to nearly 39,000 people in only 2 km², Monaco is the most densely populated country in the world (Fig. 5.3). Native Monégasques comprise less than one third of the country's population (32%). The remaining residents are French nationals (20%), Italians (15%), and a significant mix of Europeans, British, Americans, Russians and other nationalities (Central Intelligence Agency, 2020). The primary language of Monaco is French, although a minority of its native inhabitants still speak Monégasque, an ancient Ligurian

Fig. 5.3. Overview of Monaco, the most densely populated country on earth

Fig. 5.4. The open France-Monaco border, with a border marker on the right and the Monégasque language on the welcome sign

dialect (Waldman and Mason, 2006). Italian is also widely spoken and used alongside French in everyday commerce. Roman Catholicism is the dominant religion, although there is a small Jewish and Muslim minority. Owing to its Christian heritage, there are a number of churches scattered throughout the country, and a variety of religious events are celebrated each year.

Neighborly Relations

Through bilateral treaties, the international border that separates Monaco and France is entirely open to the flow of people and goods (Fig. 5.4). Passport inspections and customs controls have not existed since the nineteenth century, although France closed the border briefly in a 1962–1963 dispute about French citizens living in the principality without paying French taxes (Bézias, 2007). A 1918 treaty established a special relationship between France and Monaco, enabling France to assist Monaco in most of its international endeavors and diplomatic relationships. Another treaty in 1963 effectively established a full customs union between the two countries, and Monaco's external EU border (i.e. the Mediterranean Sea) is patrolled by the French, including passport and customs inspections on ships, yachts and helicopters arriving from outside the European Union.

The principality maintains diplomatic relationships with many other countries and hosts a French and an Italian embassy, as well as more than 70 honorary consulates within its borders. Most countries' diplomatic interests with Monaco are handled by their Embassies in Paris or consulates in other French cities. Monaco maintains full embassies in Belgium, France, Germany, Italy, Portugal, Spain, Switzerland, the UK, the USA, and the Vatican City (Holy See).

Outside these ten countries, France's embassies manage Monacan affairs.

Monaco is a full member of the United Nations and Council of Europe. However, it is not a member of the European Union or signatory to the Schengen Agreement. This fact notwithstanding, through its special relationship with France, it is a de facto participant in the European Union and Schengen Area and enjoys nearly all the benefits of full membership (Bartmann, 2012; Grinda, 2006). 'Although Monaco officially lies outside the European Union and the Schengen Zone, in all respects it operates functionally as part of both through its close relationship, open border and formal agreements with France' and the European Union (Timothy, 2017d, p. 839), and its legal currency is the euro. By agreement with the EU, Monaco has minted its own euro coins since 2001.

As a UN member, Monaco maintains diplomatic relations with most other sovereign states, but because of its small size, there are only three foreign embassies within the borders of the principality: France, SMOM and Italy. For its small size, however, as noted above, it surprisingly hosts dozens of honorary consulates from all around the world. Most countries' emissaries in Paris are double-accredited to Monaco. While most of its diplomatic responsibilities abroad are shouldered by French ambassadors, the principality has established full embassies in nine European countries (Fig. 5.5), one in Washington, DC (which is also accredited to Canada), and a full consulate-general in New York City (Embassy of Monaco, 2020).

Economy

Although there is a notable manufacturing sector (e.g. plastic and rubber products, chemicals, clothing and food products), and a thriving technology and science industry, service industries dominate (86%) the national economy. Of these,

Fig. 5.5. The Embassy of Monaco to the Vatican and Sovereign Military Order of Malta (SMOM)

banking and insurance, real estate sales, and tourism are the most prevalent (Government of Monaco, 2020). The required labor force in the principality is more than 52,000, which means that the local population is insufficient to fill all work requirements. Thus, many people commute for work daily from France and Italy. Monaco has one of the lowest unemployment rates in the world, and the Monégasques have a very high standard of living. According to the most recent United Nations (2020) data, Monaco has the world highest GDP per capita at $185,835 USD (2018 figures).

Tourism and Transportation in Monaco

Tourism plays an extremely important role in Monaco's economic and social life. It employs thousands of people, most of them being cross-border workers or foreign residents. Hotels, food services, casino patronage, yacht docking, second home ownership (real estate sector), and special events comprise a significant part of the national economy (Government of Monaco, 2020; Spence, 2016). Overall, tourism in Monaco is characterized by high-end, luxury tourism with a focus on multi-million-dollar apartments (first and second homes) and yachting (including several yacht dealerships), as well as casino gambling, luxury shopping, cruises, and a jet set lifestyle in general. As one Monaco-produced brochure points out, 'In the late nineteenth and early twentieth centuries, the era known as Belle Epoque, Monaco flourished. European nobility, tycoons, and world-renowned entertainers flock to Monte Carlo, lured by its reputation of epicurean delights, electrifying atmosphere and limitless luxury and sunshine' (Monaco Government Tourist Office, 2012, p. 5).

This image is often emphasized in the country's marketing campaigns and is a reputation that pervades public opinion throughout the world (Gay, 1998; Lew *et al.*, 2015; Spence, 2016; Timothy, 2017d); the principality is 'a playground of the rich, the famous and chic' (Page and Connell, 2006, p. 26). Chacko (1996, p. 74) similarly notes, it has positioned itself as 'a fairy tale that does not end at midnight'. The principality's own marketing materials stress the same position: 'In a land of excitement and glamour...pleasure can be as boundless as your imagination' (Monaco Government Tourist Office, 2012, p. 1). And, 'Long considered one of Europe's most celebrated and captivating destinations, the Principality welcomes a new generation of travelers with its singular mix of imperial tradition, rich history and Mediterranean sophistication' (Monaco Government Tourist Office, 2012, p. 12).

For the most part, Monaco has seen a growth in tourism since the early 2000s, reaching its zenith in 2010; the data show that there has been a slight decline since 2010. In common with most countries, data provided by the government of Monaco only count overnight stayers as 'tourists', and the only same-day visits Monaco officials calculate are cruise passengers (Table 5.1), because these can be recorded in the process of passenger disembarkation at the port. Because its border with France is fully open, data about other day-trippers is unavailable. Nevertheless, the cross-border day trip market is extremely lucrative for the country's overall tourism sector, and has been for many years, as it is with the other European microstates (Eccardt, 2005; Jenner and Smith, 1993; Timothy, 2017d). Given that Monaco is a popular excursion destination from Nice and other locations on the Côte d'Azur, as well as from north-western Italy and even further afield, it is likely that day visits far exceed the number of overnight stays and cruise passengers combined. Lodging is expensive in Monaco, so many tourists book rooms nearby in Cap-d'Ail or Beausoleil, France, and walk into Monaco as often as they desire.

As Table 5.2 illustrates, the French comprise nearly one quarter of Monaco's overnight foreign guests, following by Italians. This is unsurprising, given the geography and location of the principality, and the strong everyday connections with France and Italy. British and American tourists comprise the next largest visitor segments, which are usually affiliated with the casino and the real estate market. As the table shows, Monaco's market is fairly widespread and diverse.

Tourism seasonality shows an interesting pattern. For international tourists, the high season extends from April to October, with a peak

Table 5.1. Inbound tourism to Monaco

	2002	2004	2006	2008	2010	2012	2014	2016	2018
Overnight tourists	263,000	250,000	313,000	324,000	279,000	292,000	329,000	336,000	347,000
Same-day visitors	n/a	n/a	180,000	232,000	322,000	233,000	194,000	182,000	169,000
Total international arrivals	263,000	250,000	493,000	556,000	601,000	525,000	523,000	518,000	516,000

Source: World Tourism Organization (2008, 2011, 2015, 2020a)
[a]Same-day visits include only cruise passengers

Table 5.2. Primary markets for Monaco's tourism industry, 2018

Country of origin	Number of overnight tourists	Share of total tourist stays
France	87,016	23.5
Italy	49,558	13.4
UK	41,399	11.2
USA	34,772	9.4
Germany	14,673	4.0
Russia	14,251	3.9
Switzerland	13,420	3.6
Belgium	6996	1.9
Monaco	6439	1.7
Canada	5613	1.5
Australia	5537	1.5
Spain	5205	1.4
Netherlands	5050	1.4
China	4126	1.1
Ukraine	4110	1.1

Source: Government of Monaco (2020)

period in July and August. This corresponds with traditional summer holidays in Europe and other markets. However, for French and Monégasque tourists (presumably those residing abroad, although this is not confirmed in the datasets), the high season is October-March (Government of Monaco, 2020).

Although government records list the real estate market as a separate industry (see Government of Monaco, 2020), the property sector has clear and salient overlaps with tourism, just as it does in property development-focused Dubai and Qatar. To receive tax-free benefits, homebuyers must reside within the country. None the less, many investors buy dwellings as seasonal second homes, and these are a critical part of the state's tourism system (Fig. 5.6). Today, Monaco is the world's most expensive property market, with residential prices averaging 49,000 euros per square meter in 2018 (Government of Monaco, 2020). These figures are ten percent higher than prices in Hong Kong and 96% higher than New York's real estate costs (Bernouin, 2020). In the span of 9 years, between 2009 and 2018, prices nearly doubled from the 25,000 euros per square meter. In 2009, just over 200 homes (apartments and villas) were sold in Monaco, with a heavy growth in sales in 2013 and 2014. In 2014, nearly 600 dwelling units were sold, although there has been a slight decline since then. In 2018, 451 housing units were purchased (Government of Monaco, 2020).

The country's most famous attractions are the Casino de Monte Carlo and the Monaco Grand Prix race (Leiper, 1989; O'Kane, 2014; Timothy, 2017d) (Fig. 5.7). The casino was established in 1863 as a means of growing the faltering economy and saving the royal family from bankruptcy. Prince Charles III saw this as a way of capitalizing on the state's sovereignty (casino gaming was prohibited in France) and its favorable climate. Being aware of the social and individual dangers of gambling, he forbad members of his own family and citizens of Monaco from participating in the vices associated within the casino (King, 2008).

Another major asset is the Monaco Grand Prix. Since it began in 1929, the Grand Prix has brought considerable visibility to the principality and branded it one of the world's top motor sport tourism destinations (O'Kane, 2014). The renowned Formula One race is held each spring with the country's main thoroughfare doubling as the racetrack.

Monaco has become increasingly popular as a Mediterranean cruise stopover. In 2018, 171 ships docked at the port of Monaco releasing 168,273 passengers to explore the microstate. Cruise arrivals reached a peak in 2015 and have decreased between 2016 and 2019, with the exception of luxury cruises, which continue to grow since 2014 (Government of Monaco, 2020). This aligns with Monaco's reputation as a high-end destination. Ordinary cruises demonstrate notable variations in seasonality, with most stopovers occurring between April and October. Luxury cruises demonstrate less seasonal variation; they are popular year-round in Monaco.

In addition to the extreme high-end attractions and activities that Monaco is best known for, the country has a rich complex of cultural heritage that provides a secondary appeal for long-stay tourists and same-day visitors. Museums highlight the history of the microstate, its microstate uniqueness, and its traditional dependence on the sea (Fig. 5.8). Several theaters

Fig. 5.6. Multi-million-euro apartments are still being built in the principality to satisfy the real estate needs of the world's wealthy

are venues for the arts and entertainment, and historic buildings such as the Cathedral of Monaco, the royal Palace of Monaco, and the Palace of Justice, are important heritage attractions (Monaco Tourist and Convention Authority, 2020) (Figs 5.9 and 5.10). Also part of the local heritage, much of the country's appeal lies in its microstate status, which not only entreats curiosity-seekers but also allows certain activities that would not normally be permissible in nearby France or other parts of Europe. The casino is the best example of this, but there other border-related implications, such as income tax-free status, that contribute to the country's tourism sector. With the exception of French citizens, who must pay income taxes, there is no personal income tax for residents, and business taxes are minimal. This has turned the tiny state into a tax-free home base for many of Europe's wealthiest and has made the country a growing corporate center for global businesses.

Whereas Monaco is a luxury shopping destination, it does not have the same appeal of ordinary cross-border shopping that the other microstates have. Although it is not an EU member, it follows EU VAT tax rules and has adopted the same 20% VAT rate as France. In this way, the principality has been able to eschew the low-end, ordinary cross-border shopping that is so common in many border areas throughout the world. Besides not having room for large shopping centers, avoiding ordinary shopping tourism helps the country maintain its elite global image.

Because the principality is so crowded and lacks space, the country's handful of gardens and parks have become popular attractions for tourists, but they are especially important leisure spaces for residents (Fig. 5.11). As well, leisure park-like spaces have even been created on top of high-rise apartment buildings to satiate the needs of Monacans to have leisure spaces. These provide coveted open space in the

Fig. 5.7. The Casino de Monte Carlo is one of Monaco’s signature tourism icons and heritage sites

Fig. 5.8. The stamp and currency museum documents the microstate's philatelic and numismatic heritage

Fig. 5.9. The Palace of Justice is one of Monaco's most scenic historic buildings

Fig. 5.10. The prince's palace is an iconic heritage attraction in the microstate

Fig. 5.11. The Princess Grace Rose Garden is one of relatively few recreational green spaces in Monaco

most crowded country in the world. In accordance with Monaco's EU agreements, Monacan planners must abide by EU urban development policies, one of which requires open space to be created when natural areas are used for new construction (d'Hauteserre, 2001, 2005), although it is unclear how land reclamation plays into the regulation since it likely does not meet the standard definition of 'natural areas'.

Monaco is easily accessible from France and Italy. The French Riviera's vast coastal road network connects to the tiny country in several locations. There are two main road crossings on the western border and three on the eastern border, with many secondary streets passing over the international boundary along its entire 5.5 km length. Monaco itself has a vast network of avenues and boulevards that pervades all neighborhoods in the country. In fact, Monaco has the distinction of being the most paved country in the world. The principality has the world's highest road density—3850 km of road per 100 square kilometers of area (KNOEMA, 2020). Using this expansive road network are frequent bus connections to many communities in France and Italy, and the national bus company, *La Compagnie des Autobus de Monaco*, provides transportation throughout the country on several different lines. Rideshare companies, such as Uber and Lyft, are not allowed to operate in Monaco. French rideshare cars can drop off passengers in the principality, but they cannot pick passengers up.

Monaco has a train station but no rail services of its own. The French rail system provides train connections to Nice and Menton, and international services provide connections between Monaco and points in northern Italy. The French-Monégasque international border divides the train station, and most of the station and the country's entire railway are underground. The nearest large airport is in Nice, France, some 35 km from the principality; most foreign tourists arrive at the Nice Côte d'Azur International Airport and transfer by shuttle bus or train. Monaco has a heliport, which is currently the headquarters of two

Fig. 5.12. The Monaco Heliport is popular for commuters from Nice and regional sightseeing flights

commercial helicopter companies: MONACAIR and Heli Air Monaco (Fig. 5.12). Both companies have regularly scheduled flights to cities in France, Switzerland and Italy, and they offer sightseeing flights over Monaco and along the Côte d'Azur. Their main product, however, is providing a seven-minute shuttle service between the Nice airport and Monaco for the country's wealthy residents and visitors.

Unique among Europe's microstates in terms of accessibility is Monaco's extensive network of stairs, lifts and escalators. Being located on the edge of the Maritime Alps system, the whole country is located on a steep slope leading to Mont Agel. In a distance of only a few hundred meters, the country's elevation inclines from sea level to 161 meters. Given the small size of the country and its crowded urban conditions, most tourists walk. Thus, except for the immediate coastal area of the harbors and beaches, visitors and residents must contend with sheer physical gradients to get from ward to ward or between attractions. Stairs have been an important part of getting around Monaco for many years, and in recent decades, the government has invested significant sums of public funds to improve its vertical infrastructure by installing many elevators and escalators to make the country more mobile.

Hotels and resorts are the primary lodging option in Monaco. At present, 12 hotels provide 2387 rooms. In 2018, the accommodations sector experienced a 66.6% occupancy rate, with a national average of 2.5 nights per person in the principality (Government of Monaco, 2020). There are many Airbnb options inside Monaco, as well as in the bordering French communities.

Outbound travel from Monaco

Being the second smallest country in the world means that even many daily errands entail crossing an international border. Several of the principality's roads weave in and out of France, and most small road frontier crossings

Table 5.3. Outbound travel by Monacans/Monégasque

	2000	2003	2006	2009	2012	2015	2018
Number of arrivals of Monacan citizens or residents in foreign countries	973,362	566,144	276,362	533,783	389,557	244,303	213,175
Number of Monacan citizens or residents staying at least one night abroad	71	186	823	1666	1956	5228	4705

Source: World Tourism Organization (2020b)

are unmarked. Thus, international mobility is an everyday occurrence for most residents of Monaco. However, these crossings are not measurable or necessary to measure.

The World Tourism Organization's data indicate that some countries record Monégasque nationals and residents crossing their borders or staying in their hotels. However, there are many inconsistencies in these data, such as some countries including Monégasque citizens and Andorrans as French arrivals. None the less, the data in Table 5.3 provide a basic understanding of some Monacan outbound tourism.

Obviously, Monacans visit France more than any other country, but France does not keep track of Monégasque border crossings or overnight stays. Italy has tried to record Monacan arrivals at national borders, but given that France and Italy share an open Schengen border it is unclear how they count Monacan frontier crossings, which they estimated to be 202,749 in 2018. Residents of the principality travel all over Europe, and despite not being EU citizens, have free access to the Schengen area without the need for documentation and border controls.

Tourism issues and challenges

In common with other microstates, Monaco depends overwhelmingly on foreign workers to staff its hospitality and tourism services, because the citizen workforce is small, and most Monacans see ordinary tourism as a menial field for employment. This has widened class differences between Monégasques and non-citizen residents, although this has not been a major problem the way it has in many other countries whose service support sectors are often staffed by guest workers and immigrants.

The most pressing issue in the principality is overcrowding and potential environmental degradation. The microstate's limited territory necessarily constrains the types and extent of urban development that can occur and the types of tourism services that can be developed. As an entirely urbanized country, Monaco's only option is to undertake land reclamation projects to extend its territory further into the Mediterranean Sea, which also raises substantial concerns over its impacts on the quality of the coastal environment (Grimaldi, 2006).

Of the five sovereign microstates examined in this book, Monaco is the least independent. To preserve its monarchy and sovereignty against potential absorption by France in the 1800s and 1900s, it has had to concede many of its national freedoms to the French. This means that its laws and politico-economic policies have had to align with those of France, and its larger neighbor has traditionally taken care of the microstate's foreign affairs, including trade and border management. Monaco's alignment with France has included some tourism-related issues, such as cross-border mobility, environmental management, the use of the French franc before 2002 and the euro after 2002, and multilateral trade. Exceptions where Monégasque laws differ from those of France include casino gambling, which was grandfathered in, and the non-taxation of non-French citizens who reside in the principality.

Conclusion

Of the five microstates examined in this book, Monaco is the most heavily dependent upon its larger neighbor. Almost all of its authority to negotiate international treaties and secure its state borders has been deputed to France, which has caused some countries in the UN to question its legitimacy as a sovereign state, although it was admitted with full membership in 1993. Although France has strong-armed the principality in the past to ensure its own interests, Monaco maintains sovereignty over its internal affairs. So, despite its close relationship with France, Monaco is sovereign enough to have been able to develop its own tourism policies, such as the famed casino, its luxury tourism focus, the development of high-end real estate, and significant tax breaks for some of the world's wealthiest elites.

The principality's focus on high-end tourism has been profitable. It requires many international day workers to commute from France and Italy to staff local tourism and hospitality positions. This singular focus precludes much of the ordinary types of tourism found in other Mediterranean destinations, but because of its success and territorial limitations, the country's officials are content with their current market.

6

More Than Just a Mountain: The Republic of San Marino

Introduction

As the title of this chapter suggests, there is a pervasive, albeit inaccurate, sentiment, often exacerbated by the media and travel journalists, that San Marino is a mountaintop republic. Truth is, Mount Titano played a critical role in the microstate's historical development, but the capital city atop the mountain is only as small portion of the country's total geography and only part of what creates its tourism appeal. San Marino has a significant rural hinterland that is home to many villages, vineyards and agricultural areas, rural routes, historic buildings and contemporary industrial complexes.

This tiny republic is so interconnected with the state that surrounds it, that many tourists and travel companies mistake it for an Italian city on a hill. The Sammarinese are a proud people who abhor the idea of being confused for Italians. However, with its open borders, common language and similar architecture, un-savvy tourists might be forgiven for making this common mistake. San Marino claims to be the oldest republic in the world, and for that distinction alone, its citizens are proud to be unique. Tourism is a crucial part of the country's social life and economy. This chapter examines the unique distinguishing characteristics of San Marino, as well as its international relations and how its statehood manifests in tourism.

Place and Time

The Republic of San Marino (RSM) is the fifth smallest country in the world both in area (61 km^2) and in population (34,232). The whole state is enclosed by Italy near the Adriatic coast, bordering the Italian regions of Marche and Emilia-Romagna (Fig. 6.1). The capital city, San Marino City (*Città di San Marino*), is located near the center of the country atop Mount Titano. San Marino is not a city-state, although it was several centuries ago. Today, it encompasses a rather large rural area surrounding Mount Titano (Hamilton, 1923). The state is divided administratively into nine municipalities (*castelli*): Acquaviva, Borgo Maggiore, Chiesanuova, Domagnano, Faetano, Fiorentino, Montegiardino, San Marino, and Serravalle. Each of these has at least one urban settlement and a rural hinterland.

San Marino claims to be the world's oldest republic, with its independence having been maintained since 301 AD. Although its early

Fig. 6.1. San Marino is entirely enclaved by its large neighbor, Italy (Source: Central Intelligence Agency, 2020)

history is unclear, according to legend, Saint Marinus, a refugee from Croatia, fled to Mount Titano where he built a monastery and chapel. The hermitage later grew into a small community of religious devotees. Historic evidence from the ninth century suggests that a monastic village did inhabit the top of Titano and by the thirteen century, it was well established and under considerable pressure by the Catholic Church to become annexed into the Papal States (Cardinali, 1982). Nevertheless, it was successful in staying off the efforts of the Holy See and maintaining its independence from the 1290s until the present day. For centuries, the territory of San Marino only included Mount Titano. However, in the fourteenth and fifteenth centuries, the other *castelli* joined or were annexed into the republic.

Since its foundation, San Marino has withstood external efforts at assimilation, occupation, and political and economic incentives to relinquish its sovereignty. The country has two heads of state who serve simultaneously: the Captains Regent. The 60-member legislative body is known as the Grand and General Council (Eccardt, 2005), and even though San Marino is a unitary state, each municipality has its own nine-member council to deal with local affairs.

The population of San Marino is comprised primarily of Sammarinese nationals, with a large Italian immigrant community. The official language is Italian, and more than 99% of the population belong to the Roman Catholic Church. Although many cultural elements (e.g. language and cuisine) resemble those of the Italian regions that surround it, San Marino's inhabitants are extremely proud of their unique identity and tend to showcase their own versions of the Italian language and culinary heritage.

Neighborly Relations

San Marino has an extremely close relationship with Italy. They collaborate in many areas, including law enforcement in relation to tax evasion and smuggling (Prati *et al.*, 1996). Although the microstate's Secretariat of State for Justice oversees the legal system, through bilateral agreements most of San Marino's judges and many of its police force are required to be foreigners, primarily Italians (Cohen, 1993), whose services are subcontracted by the state as a means of avoiding nepotism and favoritism. The Council of Twelve, the country's highest tribunal and final arbiter comprises Sammarinese citizens appointed by the Grand and General Council. The country's prisoners are incarcerated in Italian jails (World Prison Brief, 2020).

San Marino is not a signatory of the Schengen Agreement or a member of the European Union. However, through earlier bilateral agreements with Italy, San Marino's border remains open, so it functions as though it were a Schengen state for cross-border traffic (Fig. 6.2). Occasionally, however, the Sammarinese police set up frontier checkpoints for law enforcement purposes and to check identity documents. San Marino has signed a customs union treaty with the EU, which allows the country to participate in the European Common Market to eliminate trade barriers between itself and other member states (i.e. Italy). However, it is not an EU member and is therefore not obligated to follow all the regulations of the European Common Market regarding imports, exports and taxation. In 2001, RSM received the right to mint its own euro coins and use the euro as its national currency.

Fig. 6.2. San Marino's open border with Italy can be crossed without passports or visas

A fully recognized sovereign state, San Marino is a member of the United Nations and its affiliated organizations. It maintains ten embassies abroad: Indonesia, Japan, the USA, the Vatican (Holy See), Spain, France, Bosnia and Herzegovina, Belgium, Austria and Italy (Fig. 6.3). Three foreign embassies are located inside San Marino City (Holy See, Italy and SMOM). There are several honorary consulates located in the country, but most foreign embassies in Rome are accredited to San Marino.

Economy

With a per capita GDP of $48,473 USD in 2018, somewhat higher than Italy's per capita GDP of $34,389 USD the same year (United Nations, 2020), the Sammarinese standard of living is relatively high. Tourism is the country's largest single industry, comprising nearly a quarter of its economy, followed by banking and manufacturing (electronics, ceramics, tiles, building material, clothing and fabric, paint and furniture). There is also a thriving agricultural sector, with the main products being grain, wine and cheese.

Despite lower demand from Italy for Sammarinese products and a weakened economy in the period 2008–2010, government incentives, such as low corporate taxes and minimal taxes on interest earnings, in the decade since 2010 there has been a rapid growth in the manufacturing sector as companies have relocated to the republic (International Business Publications, 2019). Much of the physical growth of industry has taken place adjacent to the country's border with Italy, which has traditionally accounted for some 90% of RSM's export market (Central Intelligence Agency, 2020).

Tourism and Transportation in San Marino

As noted above, tourism is the republic's most valuable economic activity and has been a

Fig. 6.3. San Marino's embassy to Italy is located in a residential neighborhood of Rome

critical development tool since at least the early twentieth century (Hamilton, 1923; Timothy, 2017e). There has been a steady growth in visitors since the 1950s. In 2018, 84,000 foreign tourists spent at least one night in the country (Table 6.1). However, the foundation of San Marino's tourism industry, as it is with most of the continental microstates, is same-day visits from neighboring countries (Jenner and Smith, 1993; Timothy, 2001b, 2017e). In 2018, 1.79 million day visitors were estimated to have arrived in RSM, compared to 84,000 overnight guests (see Table 6.1). The highest number of day-trippers are Italians, but the republic is also fashionable among same-day tourists from other countries who are in the area visiting Rimini and Bologna. Rimini is a popular coastal resort area, and Bologna is a prevalent heritage destination. San Marino is a convenient same-day journey from both of these tourism centers.

As expected, the importance of San Marino's arrivals from Italy is enormous (Table 6.2). Italians comprise 67% of international arrivals, followed by Germany (6.62%), Russia (5.48%), Poland (3.15%), France (1.88%) and Belgium

Table 6.1. Inbound tourism to San Marino (millions)

	2002	2004	2006	2008	2010	2012	2014	2016	2018
Overnight tourists	n/a	n/a	0.050	0.115	0.120	0.139	0.075	0.060	0.084
Same-day visitors	n/a	n/a	2.086	1.997	1.856	1.730	1.929	1.940	1.790
Total international arrivals	3.102	2.812	2.136	2.112	1.976	1.869	2.004	2.000	1.874

Source: World Tourism Organization (2008, 2011, 2015, 2020a)

Table 6.2. Top tourist markets in San Marino, by arriving nationals, 2018

Country of origin	Number of arrivals	Share of total arrivals
Italy	1,252,656	66.84%
Germany	124,001	6.62%
Russia	102,648	5.48%
Poland	59,029	3.15%
France	35,230	1.88%
Belgium	21,532	1.15%

Source: (World Tourism Organization, 2020a)

(1.15%). Other important markets include the Czech Republic, Switzerland, the Netherlands and China. The growing Chinese presence is not surprising given the rapid growth of Chinese travel to all Mediterranean countries, including Spain and Italy, so a side trip to RSM makes considerable sense.

In 2018, the country was home to 43 official lodging establishments and 178 food service providers. At the time of writing, there were also 52 Airbnb rentals available. The average length of stay for overnight tourists was 1.81 nights, and approximately 3500 people were employed in tourism (World Tourism Organization, 2020a), including a significant number of tourism-employed Italians who commute daily from Rimini.

The republic does not have a commercial airport. The nearest airports are Bologna's Guglielmo Marconi International Airport, 135km away, and Rimini's Federico Fellini International Airport. There is a small airstrip for personal and hobby aircraft use in the village of Torraccia, as well as a helipad. The country's commercial heliport is located in Borgo Maggiore.

In 1932, an international railway opened between Rimini and San Marino City, but it was severely damaged during the Second World War and has never been repaired. Much of the disused rail line functions as green space and walking and cycling trails. In 2012, a small stretch of the railway in San Marino was reopened for a short period for nostalgic purposes (Il Resto del Carlino, 2012). Discussions about restoring and reopening the Ferrovia Rimini-San Marino for tourist transportation received positive feedback, but it appears that talks have stalled, as the project has not made much headway since 2012. A 340-meter aerial cable car from Borgo Maggiore at the foot of Mount Titano transports passengers to the heart of San Marino City on the mountaintop (Cohen, 1993) (Fig. 6.4). This is an advantageous mechanism for car-based tourists, because the capital is a crowded Medieval city with narrow roads and few car parks (Fig. 6.5).

With more than 220km of roads, after Monaco and Macau (which is not a sovereign state), San Marino has the third highest road density of any other country: 584km of road per 100 square kilometers of territory. This figure encompasses all roads—motorways, highways, main or national roads, secondary and regional roads, and other rural and urban roads (KNOEMA, 2020). Uber, Lyft and other rideshare companies do not operate in San Marino or in nearby Italian cities. The republic is well connected to coastal Rimini, with regular bus services between San Marino City and Rimini every day. Most overnight tourists and day visitors arrive in San Marino by bus from Rimini or by private car. As such, the majority of foreign guests only visit the mountaintop capital city. Relatively few tourists visit the other eight municipalities or the countryside, but there is a national bus system that transports passengers throughout the country, stopping in most towns and settlements.

The most prominent type of tourism in San Marino is heritage tourism. This takes many different forms based on the country's cultural assets. The capital city was designated a UNESCO World Heritage Site in 2008 as San Marino Historic Centre and Mount Titano (Fig. 6.6). It was inscribed because the settlement and its mountaintop location are one of the best examples of the early Italian city-states, its eleventh and thirteenth century fortresses, and its intense collection of medieval buildings (UNESCO, 2020) (Fig. 6.7). *Città di San Marino* is the most visited place in the country, and relatively few tourists (other than Italians) venture beyond the capital to explore other communities or rural areas (Eccardt, 2005; Timothy, 2017e). The country is home to several museums, including a philatelic museum, which is extremely popular among stamp collectors. The stamps and coins of the republic are prized by collectors throughout the world.

Fig. 6.4. The Mount Titano cable car provides easy access to the mountaintop capital city

Fig. 6.5. The streets of San Marino City are narrow and usually crowded during peak tourist season

Fig. 6.6. A view of Mount Titano from the Sammarinese countryside

Fig. 6.7. San Marino's small parliament building is one of its most important heritage attractions

The Sammarinese are extremely proud of their independent nationhood and are deeply concerned about being confused with Italians (Cohen, 1993; Timothy, 2017e; Waldman and Mason, 2006). Their national pride is reflected by the fact that they have a small selection of Sammarinese culinary dishes that differ slightly from the Italian cuisine that surrounds them. Local gastronomy, their ability to maintain their independence for centuries and the very local dialect of Italian spoken by some citizens, are expressed points of pride throughout the country. Local culinary traditions are an important manifestation of Sammarinese heritage, and so are the country's wines. The burgeoning wine industry has started to create demand for rural wine tourism among car-based visitors.

In addition to heritage, business travel is emerging as an important type of tourism as well. Official data show that 30% of all arrivals in San Marino in 2014 and 25% in 2015 came for business purposes (World Tourism Organization, 2020a). The boom in industrial development within the state's borders is the most likely reasons for the importance of business tourism, as

Table 6.3. Outbound travel by Sammarinese

	2000	2003	2006	2009	2012	2015	2018
Number of arrivals of Sammarinese citizens or residents in foreign countries	1752	2263	14,754	27,681	16,580	42,348	3489
Number of Sammarinese citizens or residents staying at least one night abroad	185	244	225	392	1268	1297	2121

Source: World Tourism Organization (2020b)

well as a few conference centers now providing space for congresses and conferences (Timothy, 2017e).

Shopping is another popular activity that draws Italians and others into the country. San Marino does not participate fully in the EU's VAT (value added tax) area, so its VAT taxes are a few percentage points lower than in Italy. As well, San Marino is a tax-free retail destination for foreign tourists, not including Italians. Tobacco products and other highly taxed items in the EU are less expensive in San Marino, and the country is home to many boutique shops, outlet stores and several shopping centers.

Outbound travel from San Marino

Like residents in the other four European mini microstates, Sammarinese travel abroad frequently. Because of its proximity to, and enclaved geography within, Italy, that is the most visited foreign country. However, like other Europeans, Sammarinese regularly visit countries throughout the world. Table 6.3 provides some basic data about San Marino citizens' or residents' volume of international travel. However, the data in this table need some explanation. First, some countries count Sammarinese nationals as Italian citizens in their immigration records, and other countries only began counting Sammarinese citizens late in the 2000s. Second, Italy only began counting visits from San Marino in 2004 and ceased reporting arrivals from the microstate in 2016. The pre-2018 figures are obviously grossly underestimated, and the boundary-crossing data by Italy is inconsistent. Third, it is unclear how Italy recorded Sammarinese arrivals, since there are no formal checkpoints at the state's borders. Finally, overnight tourists and day-trippers are sometimes double counted.

Tourism issues and concerns

Overcrowding is a salient concern on Mount Titano in the capital city during the country's high tourist season (Fig. 6.8). The small capital is already crowded with narrow streets, medieval buildings and tourists, and it is becoming an increasingly popular destination, especially among day visitors. Many of these same-day excursionists spend very little money in San Marino, but they contribute to the costs of tourism, including crowdedness and socio-cultural degradation. Parking is scarce (Fig. 6.9) and San Marino City's streets accommodate large numbers of vehicles and have therefore largely been closed off to vehicular traffic.

Like Liechtenstein, San Marino suffers from a lack of international visibility, and when non-Europeans visit, they often mistake it for an ancient village in Italy. Similarly, tourists usually only visit the capital city on Mount Titano, again leading them to a misconception that San Marino is a city-state without a rural hinterland, even though there are many rural villages and cultural assets to visit outside the capital.

San Marino suffers the same fate as Andorra. It is relatively isolated, and access is easiest by personal car and inter-urban bus. This differs from the situations in the more accessible Monaco, Liechtenstein and Vatican City. The heavy reliance on coach transport from Rimini to RSM also contributes to the misperception that RSM is synonymous with its mountaintop capital, as this is the main dropoff point for the Rimini buses.

Fig. 6.8. The historic capital experiences overtourism during the summer months

Fig. 6.9. One of the few car parks available in the capital city

Conclusion

San Marino suffers a shared problem with the other microstates: it has image difficulties. It does not have a negative image; it only lacks an image on a global level. As Monaco is often mistaken as part of France, most tourists are unware that the Vatican is not part of Italy, and Liechtenstein is frequently thought to be part of Switzerland, RSM is commonly viewed as just another Italian town. The Sammarinese are extremely proud of their unique and separate identity; although they are Italian in ethnicity, they are not Italians, which is all the more frustrating because many countries enumerate them as Italian citizens when they travel. The other misperception of RSM is that the capital city on Mount Titano is the extent of the microstate's territory. Many advertisements and even travel journalists use epithets such as 'mountaintop republic' or 'the country on the mountain'. Current efforts to promote wine tourism and rural and agritourism in RSM are in part an effort to mitigate that misperception.

Small geographic area and separateness have been a benefit for RSM. Shopping opportunities have developed with lower prices on some items. Their solidarity as Sammarinese, not Italians, has created a strong sense of cultural identity that manifests in some areas of cuisine and dialect. The Sammarinese are proud to share their differences with outsiders.

In common with Vatican City, Monaco, Liechtenstein and Andorra, the primary tourism market are day visitors from nearby regions of the protector state—in this case, Italians. Day visits from the nearby coastal resort town of Rimini during high season create extremely crowded conditions in the capital city, which has been known to deteriorate the satisfaction levels of some tourists in San Marino City. Nascent discussions are under way regarding what measures might be taken to alleviate some of the overtourism the capital is beginning to feel. Another mini microstate problem facing RSM is how to develop a competitive advantage, or unique selling proposition, against nearby Italian competitor destinations in Tuscany, Marche or Emilia-Romagna, which are better known and may provide a wider range of tourism resources. Despite its quaint personality, which has attracted tourists for decades, its over-reliance on Italy has continued to perpetuate the notion that it is just another Italian destination.

7

Vatican City: The Smallest Country on Earth but Probably the Most Visited

Introduction

A small token remnant of a once substantial theocratic empire on the Italian Peninsula, Vatican City is known for two things: it is the headquarters of the Roman Catholic Church and spiritual center for the world's 1.2 billion Catholics, and it is the smallest country in the world. Both of these distinguishing characteristics play a critical role in this microstate's tourism sector. It is both a pilgrimage center, one of most visited in the world, and a quaint curiosity for non-Catholic visitors. It shares an international border only with Italy, and most of its small territory is off limits to the majority of the millions of people who visit each year.

Tourist numbers in the Vatican are excessively large, incommensurate with the country's small population and geographical size. This state of less than half a square kilometer is home to some of the world's eminent masterpieces of art and sculpture, architecture, and sacred shrines. It is also home to one of the world's smallest, yet longest serving, militaries on earth: the Swiss Guard. Vatican City is also the only country on earth that is a true theocracy and whose budget is funded primarily by donations to the church. In spite of the microstate's position as the Catholic Church headquarters and most of its funding coming from the church, other than printing, banking and real estate, tourism plays an important role in funding state operations and in maintaining its separateness from the country that surrounds it. This chapter examines the special situation of Vatican City and explores its transfrontier relations, and how these affect tourism in the microstate.

Place and Time

The Vatican City is the smallest independent country and sovereign state in the world, although its sovereignty lies with the Holy See rather than in the physical state mechanism, as sovereignty normally is possessed and exercised throughout the world. The Holy See is the 'composite of the authority, jurisdiction, and sovereignty vested in and exercised by the pope and his advisors in the temporal and spiritual direction and guidance of the Roman Catholic Church throughout the world' (US Department of State, 1987, p. 3). Thus, the Holy See is an intangible, sovereign entity, rather than a physical locality; its headquarters is the Vatican City state. It is the Holy See that negotiates diplomatic ties, maintains embassies abroad (known as Apostolic Nunciatures), hosts embassies in

Fig. 7.1. Vatican City, surrounded by Rome, is the world's smallest sovereign state (Source: Central Intelligence Agency, 2020)

Rome, and establishes international treaties, not the Vatican City state. Although sovereignty belongs to the Holy See, the Vatican City is the geographical embodiment of the Holy See's sovereign authority (Toschi, 1931). Because this book deals with places above all else, rather than non-physical entities, herein the Vatican City is treated as the world's smallest sovereign state.

Besides being the headquarters of the Roman Catholic Church, the Vatican is a functioning country. It is 0.44 km^2 in area and has a current population of approximately 800, including citizens and non-citizen residents, and it is entirely enclaved within the city of Rome (Fig. 7.1). Resident non-citizens are primarily Italians, Swiss and other Europeans. There are presently around 620 Vatican City citizens, one quarter to one half of whom live abroad; many of these are diplomatic staff. Not surprisingly, the entire population is Roman Catholic. The primary spoken language is Italian, although there is a large German-speaking population among the Swiss Guards, and most documents are written and curated simultaneously in Latin.

Although the site of the Vatican City is of ancient origins, and many of its structures are hundreds of years old, the independent state has only existed since 1929. From the mid-eighth century to 1870, the pope and Catholic Church held sovereignty over a large section of the Italian Peninsula, areas known as the Papal States or State of the Church. The Papal States was an independent country that housed the temporal and spiritual domains of the pope. Throughout history, the area of today's Italy was comprised of many small, independent kingdoms, duchies and republics. However, in the 1840s, the Kingdom of Sardinia, ruled by the Savoy family, drove an effort to unite the Italian Peninsula into a unified Italian state. During the 1850s and 1860s, the majority of Italian countries had been annexed into the new Italian state.

In 1861, the Kingdom of Italy was declared by the ruling Savoy dynasty, which by 1870 had captured essentially all of the territory of the Papal States, including most of Rome. The only areas to remain under the control of the church were a couple of administrative buildings, the Papal residence, the Vatican Gardens, and St Peter's Basilica on Vatican Hill. The Italian subjugation of the Papal States devastated relations between the Italian state and the church, and they remained at a political impasse for six decades until both entities agreed to sign the Lateran Treaty in 1929. This treaty deeded the walled area of the Vatican Hill and St Peter's Square (totaling 44 hectares/109 acres) to the Holy See as its state territory, guaranteed the sovereignty and independence of the Holy See and the Vatican City, and compensated the church financially for the loss of the Papal States (Toschi, 1931).

Approximately 84% of the country is surrounded by a 12-meter wall, most of it built between the ninth and sixteenth centuries. The microstate comprises several distinct areas: St Peter's Square and St Peter's Basilica; the state administrative zone; the Vatican Gardens (including the heliport and the radio and television stations); the Vatican Museums and Sistine Chapel; the Apostolic Palace (pope's residence) and Swiss Guard Barracks; and the commercial zone. There are eight official crossing points between Italy and the city-state, including the railway gate and certain doors that are only used on rare occasions (Fig. 7.2).

Administratively, the Vatican City is a sacerdotal monarchy (US Department of State, 1987). There, the pope has a dual role: he is both the religious leader of the world's 1.2 billion Roman

Fig. 7.2. One of the main border crossings into Vatican City

Catholics, as well as the *ex officio* head of state, or more properly, the Sovereign of the State of the Vatican City (Stato della Città del Vaticano, 1979). In this position, he exercises supreme executive, judicial and legislative power. Matters of state are typically managed by a legislative body known as the Pontifical Commission, and officers in that council serve positions equivalent to prime minister, minister of public affairs, and minister of foreign affairs (US Department of State, 1987).

Like every other country in the world, the Vatican provides many public services and operates independently of Italy, for the most part (Stato della Città del Vaticano, 1979). It has its own postal system, maintains a university, mints its own coins, and has a fully staffed fire department and public works department that maintains the streets and gardens. The state operates a hospital, pharmacy, telephone network, and television and radio stations. It also has a supermarket, a small number of restaurants and cafes, printing presses, a petrol station, a police force and a small military—the Swiss Guards (Fig. 7.3). Both the Swiss Guards and the Vatican Gendarmerie provide security services, defend the state's borders, and protect the pope (McDowell, 1991).

Neighborly Relations

As any sovereign state should, in most ways the Vatican functions independently of its neighbor, Italy. There are a few notable exceptions. First, according to treaty, the Italian police guard St Peter's Square alongside Vatican Gendarmerie and enforce both Vatican and Italian law there, up to the base of the basilica's steps. Rome's police force also operates the security checkpoints for visitors entering the basilica. Beyond St Peter's Square, the Italian police have no authority. Second, most of the laws of the Vatican state are drafted to be in harmony with Italian laws, and criminal cases are tried in Italian courts.

Fig. 7.3. The Swiss Guards are not only the pope's protectors and the state's military, they are also a tourist attraction

Third, given its microscopic size and land scarcity, the 87 current embassies assigned to the Holy See are located in the city of Rome (Vatican Secretary of State, 2020), although several of them are located very near the Vatican. Finally, also because of its small size and landlocked situation, the country receives most utility services from Italy. For instance, its water and electricity supply are furnished by Rome, garbage is collected by the city of Rome and deposited at Italian landfills, and its sewerage is treated by Rome.

In accordance with the 1929 Lateran Treaty, there are no customs or passport controls entering or departing the Vatican City territory. Nevertheless, as discussed below, there are secure and strict crossing procedures in certain areas. Tourists are only permitted to visit certain parts of the country, and guests who are able to visit must possess special permission from the state.

The Vatican City (or Holy See) is not a member of the Council of Europe or the European Union, nor has it signed treaties related to the single market or customs union. This is in part because of its status as a theocratic absolute monarchy (the pope being the monarch) rather than a democracy, which does not adhere to the EU rules of popular consent. By virtue of its close and open relationship with Italy, it functions by default as a de facto member of the EU in areas of trade, cross-border mobility and security. The Vatican's main agreement(s) with the European Union in 2000 and 2010 allows the country to mint its own euro coins and use the common currency as its national monetary unit. Because it lies outside the EU, the Vatican is not party to the alliance's VAT laws. In fact, the Vatican City is exempt from import and export duties, as well as sales taxes.

Even before the Vatican City State was created in 1929, the Holy See had had diplomatic relations with various states throughout the world for hundreds of years. The Holy See's embassies abroad are known as Apostolic

Nunciatures, and its ambassadors are nuncios. The Holy See/Vatican City is not a member of the United Nations, but it is a permanent observer state, which means that it can also participate in UN agencies through ‘Permanent Observer Missions’. It participates in UNESCO, for example, through its Permanent Observer Mission, which allowed its entire territory and its extraterritorial properties to be listed as World Heritage Sites.

Although not a member of the UN, the Vatican City maintains diplomatic relationships with nearly every sovereign state through its numerous Nunciatures. Nearly half of the world’s sovereign states (87 as of 2020) support embassies to the Holy See (all located in Rome, not on Vatican soil). Several of these are accredited to both the Vatican and Italy, or the Vatican and the Sovereign Military Order of Malta (Vatican Secretary of State, 2020). Fourteen embassies in Switzerland are accredited to the Holy See/Vatican City, and the rest are accredited to the Holy See from their embassies in other European capitals.

Economy

The economy of the Vatican City is dependent on charitable donations and direct funding from the Catholic Church. However, various investments (e.g. real estate), banking and finance, printing services, sales of stamps and coins, and tourism comprise the majority of the country’s non-donated revenue. While of little economic value, there is even a small agricultural sector. A large vegetable garden provides fresh-grown sustenance for the pope and other state dignitaries, and on state lands (extraterritorial properties) at the pope’s summer residence, Castel Gandolfo, cows provide milk for use in Vatican kitchens and is sold exclusively, together with pontifical yoghurt, at the Vatican supermarket (Fig. 7.4). Besides funding from the Catholic Church, the

Fig. 7.4. Vatican milk sold only in the country’s one supermarket

state is highly dependent upon tourist spending. The Vatican City state receives a considerable sum of money each year from the sale of souvenirs, museum admission fees, tours, sales of publications, and sales of postage stamps. Morris (2019) estimates that the country earns approximately 80 million euros per annum from museum entrance fees and 20 million euros from souvenir sales and corporate hospitality. Although extremely small, the Vatican has a large workforce of approximately 5000 people who cross the border from Rome each day to work in tourism and other services (Central Intelligence Agency, 2020).

Tourism and Transportation in Vatican City

Providing for the spiritual needs of the world's Roman Catholics is obviously the main goal of the Holy See. However, Vatican City is an extremely important tourist destination, despite the fact that visitors are not permitted to wander around the country freely. Only certain areas are open to tourism; less accessible areas (e.g. the Vatican Gardens and the Necropolis) can only be visited on guided tours with prior arrangements, and most of the country cannot be visited except by residents and guests of the state or church.

Given that St Peter's Square is open to the public, and passports are not required to enter the world's smallest state, it is impossible to know how many people cross the Vatican City-Italy border each day. Although France has topped the World Tourism Organization's list of most visited destination countries for many years, it is likely that the Vatican City is in reality the most visited country, if international day trips are considered. Rome is Italy's most visited destination, and in 2018, an estimated 29 million foreign and domestic tourists visited the city (Instituto Nazionale di Statistica, 2019). Although there are no specific data to verify the assumption, it is highly likely that nearly all tourists who visit Rome also visit the city-state (Morris, 2019). Every popular guidebook, webpage, and standard tour of Rome includes visits to the Vatican Museums and St Peter's Basilica as essential elements of a Roman vacation. In addition to out-of-town tourists, hundreds of Romans cross into the Vatican each day for various purposes beyond employment: to worship and pray in St Peter's or at one of the microstate's other shrines or chapels, to shop, refill prescriptions, purchase gasoline, or visit friends and relatives who dwell within the country's walls.

The foremost type of tourism in the Vatican is religious tourism, including both pure pilgrimage and less-pious visits (Kim and Kim, 2019; Ron and Timothy, 2019; Vukonić, 2006). Rome in general has been a significant pilgrimage destination since Christianity was legalized in the Roman Empire in 313 AD and with the rise of the cult of saints in which the lives and martyrdoms of the apostles Peter and Paul were commemorated and revered (Birch, 1998). Throughout the Medieval period, pilgrimage from all over Europe to the area of today's Vatican City was relatively routine (Liutikas, 2018; Ron and Timothy, 2019; Timothy and Olsen, 2018). Because the Vatican City is the center of Roman Catholicism, the whole country is essentially sacred. None the less, there are particular areas that are more sacred than others and are visited by Catholics from around the world as part of their devotional pilgrimages.

St Peter's Basilica was constructed in various forms and at different stages between the fourth and seventeenth centuries ostensibly over the burial chamber of St Peter the Apostle (McDowell, 1991). The basilica is the sacred centerpiece of the state and its busiest tourist attraction (Fig. 7.5). The Sistine Chapel, which is home to some of the most important sacred frescoes in the world, is another key pilgrimage location (Edwards, 2018). St Peter's Square is important for Catholics, who believe it was the location of St Peter's martyrdom. It is also a key location where the pope delivers his general audience discourse each Wednesday and where millions of pilgrims gather during holy celebrations and holidays, particularly Christmas and Easter. There are several other sacred shrines and churches throughout the country, but these are generally inaccessible to the visiting public.

Religious tourism is a form of heritage tourism, but there are many other heritage and cultural assets that draw a more general tourist audience. Of particular importance in the realm of cultural heritage is that the Vatican City was inscribed on UNESCO's World Heritage List in 1984. Because of its compact size and

Fig. 7.5. St Peter's Basilica, Vatican City's most visited attraction and center of worldwide pilgrimage

high concentration of historic buildings and artworks, it is the only World Heritage Site on Earth to encompass an entire country.

The state's primary heritage attractions are St Peter's Basilica, the Vatican Museums, the Sistine Chapel, St Peter's Square, the Necropolis, and the Vatican Gardens (Fig. 7.6). As already noted, the basilica is extremely sacred for Roman Catholics. For non-pilgrim tourists it is a cultural heritage curiosity, an architectural wonder and iconic sculpture museum (Nolan and Nolan, 1992; Vukonić, 2006). It is, perhaps, the most famous church in the world and an important heritage brand for the Vatican City and the city of Rome. The Vatican Museums are home to some of the world's largest art and sculpture collections that were amassed by popes and church representatives throughout the Middle Ages and in modern history (Eccardt, 2005). The Necropolis and tomb of St Peter beneath the Basilica are an important attraction both for Catholic pilgrims and curious others, and the country's heritage in general is an important attraction for Rome's schoolchildren, who frequently undertake field trips in the gardens and historic areas. In 2003, a Roman burial ground from the first century BC to the fourth century AD was discovered during excavations for a new underground parking garage in the Vatican's commercial zone. This impressive necropolis is only accessible to a small number of visitors by appointment and is not widely promoted as a key attraction.

There are several types of tourism that are unique to the world's smallest country. The first is utilizing its efficient postal services. While this does not normally sound like a touristic activity, in the unique context of the Vatican, it is. There are three perspectives on this. First, Vatican Post (*Poste Vaticane*) is considered one of the most efficient postal systems in the world, far more reliable and faster than the *Poste Italiane* that surrounds it. As such, people who live nearby in Rome prefer to use Vatican post offices to Italian ones (Fig. 7.7). People living further away frequently use the Vatican service for mailing

Fig. 7.6. The Vatican Gardens are one of the country's prized heritage assets, but they are generally inaccessible to tourists

Fig. 7.7. Vatican City post offices play an important role in the mini state's tourism economy

packages and bulk mailings, such as Christmas cards and holiday packages. It is often said that more letters, cards and packages are mailed each year from the Vatican than from any other postal code in the world (*Telegraph*, 2016).

Another way in which postal activities are an attraction is the widespread practice of tourists mailing postcards. Although this activity has waned in the past 20 years through the increased use of social media (Munar and Jacobsen, 2014), in the Vatican, the situation is different. Many visitors, who might not normally send tangible postcards during their holidays, frequently do so from this microstate because of their novelty value. Postcards and stamps are among the most frequently purchased souvenirs in the microstate. Third, collectors generally place a high value on Vatican City stamps; they are unique and artistic, limited in number, and representative of the world's smallest country. Philatelists from around the world cherish the opportunity to visit *Poste Vaticane* and buy stamps for their collections (Timothy, 2001b).

Shopping was noted previously as an important part of other microstates' tourism industry. It is likewise a relatively important and unique commercial activity in the Vatican City, which derives from the state's independence and its border with Italy. The sorts of shopping that occur in the Vatican would not normally be considered a leisure or touristic activity. However, given the scale of the state and the relative economic importance of transfrontier retail from Rome, it should probably be considered a form of very localized utilitarian 'tourism', similar to what occurs in many other border areas throughout the world, because consumers cross an international boundary and engage in economic transactions (Timothy, 2005).

Between the north colonnade and the ancient city wall, but on Italian soil, many souvenir vendors sell a wide variety of churchy and secular mementos. Inside the Vatican boundaries, souvenirs are only sold at the four tourist-accessible post offices (philatelic and numismatic souvenirs), in the museums, at a small gift shop atop St Peter's Basilica, and at the official souvenir store in St Peter's Square. Owing to the country's tax-free economy, prices at the supermarket are significantly lower than in Rome, as is the price of gasoline at the local petrol station. The Vatican pharmacy honors Italian

Fig. 7.8. The Vatican Pharmacy is a popular place for Romans to fill their prescriptions

prescriptions (Fig. 7.8). Drug prices are significantly lower than in Italy, and certain pharmaceuticals that are not sold in Italy are available in Vatican City. These services are meant for residents, but with special permission, a small number of Italians, including but not limited to employees of Vatican institutions, are allowed to use the supermarket, pharmacy and petrol station (Fig. 7.9). Likewise, the top floors of the railway station have been converted into a tax-free department store, which only Vatican citizens and certain employees, diplomats assigned to the Holy See, and some of their dependents and family members from Rome are permitted to use.

A final unique form of tourism is a visiting friends and relatives (VFR) type of tourism. Every year, a new intake of freshly trained Swiss Guards (usually 30–35) arrive at the Vatican to be sworn into service by the pope. The Swiss Guards themselves are an important part of the heritage milieu of the Vatican and are an important attraction for tourists because of their ornate uniforms, public visibility, and unique history (see Fig. 7.3). They are required to be unmarried males, Catholic, Swiss citizens trained by the Swiss Armed Forces, and between the

Fig. 7.9. The Vatican petrol station sells gasoline at tax-free prices, but not everyone is permitted to use it

ages of 19 and 30 years. These are essentially the state's military. In the final week or so before their service begins on May 6 each year, their families are invited from Switzerland to tour the entire Vatican state with their sons acting as personal guides (Fig. 7.10).

Transportation and parking in the Vatican City are not for tourist use. There is an intricate network of roads and footpaths throughout the country, which are for local traffic, public utility services, and open bus garden tours. Due to the country's limited geography, parking is a significant concern, although there are several large car parks, including one underground, for residents and workers. Of the eight border entrances into the Vatican, standard tourists use only three of them. Uber operates in Rome, so it is possible for tourists to be dropped off by a rideshare at the Vatican City border.

The 862-meter state railway is the shortest national railroad in the world, and only some 230 meters of it are actually located on Vatican soil (Fig. 7.11). The national railway is used almost exclusively to receive supplies and export religious printed materials, although in 2014, a regular Saturday passenger service was inaugurated to transport tourists between the Vatican train station and the Pontifical villas outside of Rome. The nearest international airport is Rome's Leonardo da Vinci-Fiumicino Airport. Vatican City has a heliport, which services some of the travel needs of the pope, other Vatican officials and church leaders, as well as visiting foreign dignitaries, but does not play a traditional tourism role as the heliports in Liechtenstein and Monaco do.

There is no overnight lodging for the general public, although cardinals, bishops and other state and church dignitaries may use Martha's Guest House when they visit. Hundreds of lodging facilities, including Airbnb rentals, are available in Rome very close to the Vatican City. There is a small food services sector in the Vatican City, including a restaurant for residents only (and

Fig. 7.10. New Swiss Guards guiding their families through the city-state

Fig. 7.11. The Vatican railway and train station, which is now a duty-free shop

Table 7.1. Outbound travel by Vatican City residents and citizens

	2005	2010	2015	2018
Number of arrivals of Vatican citizens or residents in foreign countries	815	568	879	1021
Number of Vatican citizens or residents staying at least one night abroad	36	237	319	122

Source: World Tourism Organization (2010, 2014, 2020d)

Romans with permission) in the commercial zone. Tourism catering includes six eateries—cafés, pizzerias and restaurants—in the Vatican Museums and a refreshment service atop St Peter's Basilica near the dome.

Outbound travel from Vatican City

Most residents of Vatican City cross an international boundary almost daily. Many also travel further afield on state and church business, for pilgrimage purposes, as well as for pleasure. Table 7.1 shows the number of Vatican citizens and/or residents who arrived in other countries between 2005 and 2018. Italy, of course, does not tabulate visits from the Vatican, and many other countries count Vatican passport holders under the category of Italian citizens. Thus, the numbers presented in the table are grossly inaccurate, but they do suggest that the number of Vatican outbound travel is equal to, or exceeds, the number of the microstate's residents, indicated multiple journeys by at least some of the country's residents. The most visited countries outside of Italy are Israel, Poland, Jordan and Lebanon. Israel and Jordan are not surprising, given the Holy Land pilgrimage motive that might be a major driver for those who live within the Vatican walls. There is a strong Catholic connection to Lebanon, which is also considered part of the broader Holy Land. Poland is one of the most devoted Catholic countries in Europe and the original home to the much-loved former Pope John Paul II. There continues to be a strong connection between the Holy See and Poland.

Tourism issues and concerns

The church has long tried to differentiate between pilgrims and tourists, as have many scholars and religious organizations, suggesting that pilgrims are motivated to travel by spiritual needs and religious obligations, while tourists travel for hedonic and pleasurable purposes. Such a binary view suggests a lack of understanding about who is a tourist and what tourism is. According to the World Tourism Organization, a tourist is someone who travels away from home for at least one night but not more than a year, and the person should not be remunerated financially from the destination visited. One's motivation for traveling (i.e. pleasure, learning, spiritual growth, or religious obligation) is irrelevant to the definition of a tourist. Thus, pilgrimage is a niche form of tourism, and pilgrims are a niche tourist market (Ron and Timothy, 2019).

In recent years, the Catholic Church has begun to recognize and accept this fact. For more than a decade, the church has sponsored several academic and industry conferences with a focus on religious tourism to create a better understanding of this niche market that makes up such a large part of the Vatican's visitor audience. In line with this broader perspective, the Vatican operates a tour company that sells packaged pilgrimages to a variety of sacred and secular destinations, often combining the two in a single experience (Ron and Timothy, 2019). These evolving perspectives on tourism have created a more hospitable environment within the microstate, as leaders have come to realize that not everyone who visits does so for strictly spiritual reasons. Even those who visit as devout pilgrims have begun to see a more service-oriented approach to pastoral care.

Fig. 7.12. The crowded corridors of the Vatican Museums are difficult to manage and navigate

Another important issue is overtourism. Although this is not as salient an issue for the other European microstates (except San Marino's capital city), it certainly is a major concern for the Vatican City, it being the smallest country in the world but one of the most visited destinations in Europe. The Vatican Museums, the Sistine Chapel, and St Peter's Square and Basilica are extremely crowded during high season and holidays (approximately 30,000 people per day), so much so that the wait to enter the church and museums can last hours. As well, the vast crowds affect the sacredness of the Sistine Chapel. The spirit of that sacred edifice is persistently interrupted by boisterous tourists, and the museum's long corridors are crowded to the point where it is sometimes hard to move (Fig. 7.12). These conditions are exacerbated by cumbersome security checks, which are necessary due to the open border at St Peter's Square and credible terrorist threats against the state. For the protection of the artworks, employees and tourists, state leaders and museum managers are presently considering limiting the number of people who can enter the museums and chapel each day (Morris, 2019).

Conclusion

Vatican City has an extremely close relationship with Italy. Much of this relates to the threefold role of the pope: he is the worldwide leader of the Roman Catholic Church and the Bishop of Rome. His third role is more political; he is also the head of state of Vatican City and the single holder of sovereignty over the Vatican and the Holy See. In all three of these ways, the Vatican state is deeply connected to Italy, and according to the Lateran Treaty, Italy provides most services required by this sovereign microstate. Geopolitically, perhaps the most unique characteristic of the Holy See/Vatican City, besides its microscopic size, is that it is free and able to conduct all of its own international agreements and establish its own foreign missions abroad, a privilege that is limited among the other four microstates.

The hardest European microstate to study is Vatican City, because it lacks even baseline tourism data, and whatever data the state collects, it is usually unwilling to share. For historical and cultural research, its archives and libraries are extremely hard to access. The Vatican City state does not normally see itself as being in the tourism business and hence does not see the need for tourism data. However, the state and the church's perspective on tourism has begun to evolve in recent years with a broader realization that pastoral care may not be only about preaching the gospel or caring for the world's poor. It also includes providing opportunities for tourists to appreciate great works of art and architecture, and commercial activities, such as selling souvenirs, stamps and coins, guided tours, and train rides.

Because of its extraordinarily small size, ordinary commercial activities that might not be considered 'tourism' in normal situations can be counted as a microstate form of tourism here (e.g. visits to the pharmacy, the petrol station, duty free shop, and grocery store). This is in addition to the millions of visitors who come to enjoy the country's arts and architecture each year. Given its position in Rome, its popularity as a premier destination, and its constant flow of border crossings, Vatican City is probably the most visited country on earth.

8

Europe's Autonomous Regions, Dependencies and Overseas Territories

Introduction

This chapter highlights conditions associated with three related types of polities: autonomous regions, crown dependencies and overseas territories (Fig. 8.1). As Chapter 1 clarified, several European countries (e.g. France, Netherlands, Denmark, Portugal, Spain and the United Kingdom) maintain sovereignty over 40 possessions in various parts of the world, primarily in the Caribbean and Pacific Islands. These are labeled by many different names, depending on their legal classification within the state structure and how independent or autonomous they are in relation to the mother country. From the EU's perspective these are generally classified as Overseas Countries and Territories, Outermost Regions or a specialized category of their own. Although these are an extremely important part of the political geography of the European metropoles, and most are remnants of colonial empires, this book is limited to empirical discussions about polities located in Europe. Thus, the discussion on overseas territories and dependencies here concentrates on Gibraltar, the British Sovereign Base Areas, the Isle of Man, and the Channel Islands: Jersey and Guernsey.

The first part of the chapter describes the situations associated with autonomous or semi-autonomous regions in Europe. These are places, which, usually because of physical separation, isolation or a desire for secession, have a considerable degree of autonomy and freedom from external authority with regard to internal decision making and local matters. This may be seen as a devolutionary move to decentralize government authority and democratize participatory and local governance so that decisions are made by locals for their own wellbeing, according to local needs and conditions. As territories and crown dependencies of the UK, Gibraltar, the Isle of Man, Jersey and Guernsey are also simultaneously autonomous regions, which illustrates that the boundaries between these geopolitical classifications are not clear-cut.

Autonomous and Semi-Autonomous Regions

Autonomous areas are an integral part of the state, but they have a higher degree of self-rule than other areas of the state. As described above, autonomous regions have achieved this status owing to isolation, distance from the sovereign state, or by means of some sort of special negotiation between the state and the territory through an external arbiter. In the latter case, autonomy

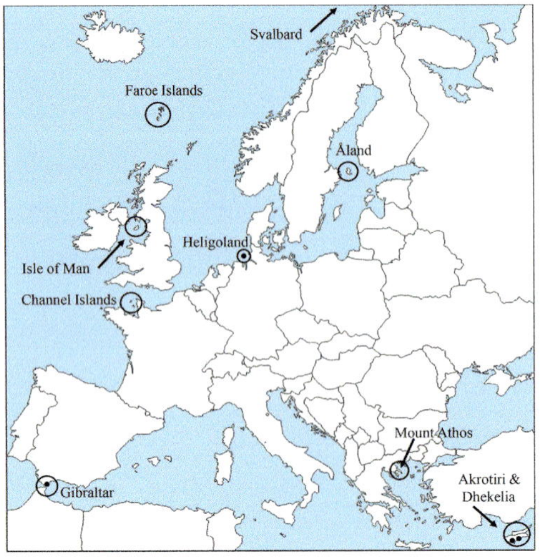

Fig. 8.1. Autonomous Regions, Overseas Territories and Crown Dependencies in Europe (Source: basemap from d-maps.com)

within a state is usually conceded to appease nations, peoples or groups that desire complete independence, greater self-determination or unification with a foreign state through irredentism. The Găgăuzia autonomous Territorial Unit (Moldova) and the Adjara Autonomous Republic (Georgia) are prime examples of this latter category.

Europe is populated by a wide range of autonomous and semi-autonomous regions (Table 8.1). The level of self-governance varies in each situation. In Greenland and the Faroe Islands, for example, home rule runs deep and enables these two constituent countries to govern almost every aspect of their daily political, economic and social machinery, with only a few things pertaining to international relations being left to Denmark (Hovgaard and Ackrén, 2017; Karlsson, 2009). Other entities, such as Heligoland, Germany, have only limited autonomy in areas of taxation and traffic regulations (Kovova *et al.*, 2018; Muller, 1977).

Given the constraints of space and the emphasis on tourism in this book, only a handful of these autonomous regions will be examined in the sections that follow. These include Heligoland (Helgoland), Åland, Mount Athos, Svalbard, and the Faroe Islands.

Heligoland

Heligoland (also spelled Helgoland) is a small two-island German archipelago 46 km off the coast of Germany. It has a population of roughly 1250 people and a thriving tourism industry. In 1890, Germany received Heligoland from the British in the Heligoland-Zanzibar Treaty. The archipelago had been a tax-free zone under the British and, as part of the 1890 agreement, that status quo would remain on the islands under German sovereignty (Muller, 1977). Today, Heligoland is a semi-autonomous part of Germany where different tax rules apply from other parts of Germany and the EU. Heligoland

Table 8.1. Examples of autonomous and semi-autonomous regions in Europe

Sovereign state	Area	Classification
Denmark	Faroe Islands	Autonomous territory
Finland	Åland	Autonomous province
Georgia	Adjara	Autonomous republic
Germany	Heligoland	Semi-autonomous area, special status area
Greece	Mount Athos	Autonomous monastic state
Italy	South Tirol	Autonomous province
Moldova	Găgăuzia	Autonomous territorial unit
Norway	Svalbard	Semi-autonomous, unincorporated area
Serbia	Vojvodina	Autonomous province
Spain	Catalonia	Autonomous community
UK	Northern Ireland	Constituent country with devolution

is part of the EU (as part of Germany), but it remains a duty-free and VAT-free area within the EU. Thus, from a fiscal perspective, it is classified as a foreign country and not subject to the EU's tax laws and regulations. As a result, as tourists depart back to the German mainland, they are required to clear German customs to ensure they do not exceed their EU customs allowances.

This tax-free advantage forms the backbone of its thriving tourism sector. Although the islands are natural and culturally attractive, and there are several inns and hotels for lodging, the majority of the Heligoland's tourists are same-day visitors from mainland Germany, who come by boat (4 hours) for the primary purpose of tax-free shopping (Gunkel, 2014; Muller, 1977). Each year approximately 250,000 day trippers, mostly Germans, visit for the purpose of buying tobacco, alcohol, perfume, electronics, clothing and other highly taxed products. This tax advantage and its touristic manifestations comprise three quarters of the islands' economy. For longer-stay visitors, the islands are home to dramatic coastal scenery, wildlife (birds and seals), walking trails, nice beaches and a nascent spa sector.

Fig. 8.2. Welcome to Åland: Autonomous and demilitarized

Åland

Finland's only autonomous region, Åland, is a Swedish-speaking archipelago of around 30,000 inhabitants that lies between Sweden and Finland in the Baltic Sea. Early in the nineteenth century, the Åland Islands became part of Finland inside the Russian Empire, and when Finland declared independence from Russia in 1917, the Ålanders felt threatened that their distinct culture and Swedish language would disappear as Finland tried to encourage the archipelago to become more Finnish (Åkermark *et al.*, 2018). At this time, the Ålanders considered themselves to be Swedish in every respect and held their own referendum on nationality, with the majority desiring to be returned to Sweden, but this did not happen. In 1921, the League of Nations concluded that Åland would remain part of Finland once Finland provided guarantees that the islands would be treated as an autonomous region with their own cultural and linguistic rights, legislative assembly and constitution, and that the islands would remain completely neutral and demilitarized (Baldacchino and Pleijel, 2010; Karlsson, 2006, 2009; Loughlin and Daftary, 1999) (Fig. 8.2).

The 1920/1921 Åland agreement was reaffirmed in 1991 when Finland negotiated its admission into the EU. Based on the results of its own referendum, Åland joined the EU with the rest of Finland in 1995, but the islands have maintained certain rights that the rest of Finland and other areas of the EU do not have. For example, although EU law requires the freedom of movement and abode within the single market, Åland retains the right to restrict the number of outsiders who can live in the archipelago or buy property there. This is in large part a means of maintaining the unique insular culture and Swedish language (Jahn, 2015; Lindström, 2000). Likewise, Åland chose to remain outside the VAT area and is exempt from certain other tax regulations.

Åland has its own elected premier and parliament, which is empowered through the Self-Governance Act to legislate separately from the Finnish state essentially all internal affairs, including important tourism-related areas such as heritage and environmental management, health care, transportation and education (see Table 8.2). The archipelago also operates its own

Table 8.2. Legislative and governance domains in the autonomous Åland Islands

Governance area	Åland's legislative rights	Finland's legislative rights
Education	X	
Culture	X	
Heritage protection	X	
Health care	X	
Environment	X	
Business development	X	
Intra-island transportation	X	
Municipal administration	X	
Police/law enforcement	X	
Postal services	X	
Radio and television	X	
Foreign service/diplomacy		X
Most civil and criminal law		X
Courts		X
Customs, passports and visas		X
National taxes		X

Source: based on information in Ålands Lagting (2020) and Loughlin & Daftary (1999)

postal system and issues its own postage stamps. According to the territory's legislative website, 'Åland operates almost as an independent state with its own legislation and administration' (Ålands Lagting, 2020, n.p.). Many of Åland's freedoms and areas of self-governance parallel the types of matters other autonomous regions also govern by themselves (see Table 8.2).

As noted above, even though Åland is part of the European Single Market with Finland, it lies outside the EU's VAT area and excise tax regulations, which means that it is classified as a third country as regards trade with other EU states (Åland Post, 2020). When the decision was made by the European Commission to eliminate tax-free sales within the EU, this would have devastating effects on Åland, which already had a robust duty-free sector on the ferry ships that plied daily between the islands and Finland and Sweden. As an autonomous region, Åland threatened to remain outside the EU unless it was permitted to continue in the tax-free retail sector. As a result, it was able to join the union while remaining outside the EU's fiscal boundaries, which enabled the province to continue its shipboard tax-free sales (Karlsson, 1999).

Tourism has long been an important economic sector in Åland, with its main markets unsurprisingly being Finland and Sweden (Baum, 1996), although only a handful of studies have examined tourism-related issues in the autonomous region in recent years (e.g. Fellman, 2016; Fellman *et al.*, 2015). The most important tourism subsector in the province is ship-based transit tourism. The islands are a stopover point on the frequent sailings between Finland and Sweden, and it has become a popular destination in its own right for day trips from Sweden and Finland (Holmberg-Anttila, 2009). These day trips entail passengers arriving, quickly changing ships at Mariehamn harbor and then returning to their starting point the same day (Forsgård, 2019). In 2018, this visitor type numbered around 1.2 million. More than 770,000 same-day visitors arrived from Stockholm the same year. For them, the journey is a 24 hour affair and includes a bit longer stopover in Åland's capital city, Mariehamn (Forsgård, 2019). Although these same-day transit visitors have little economic impact on the ground in Mariehamn, they typically spend considerable sums aboard the ships,

Fig. 8.3. Tax-free sales on the Finland-Åland ferry ships

which offer entertainment, food and tax-free sales (Fig. 8.3).

Given its Nordic and maritime location, Åland has a very short and distinct tourist season. Thus, most of its land-based tourism takes place during the summer months, whereas the ship-based tourism is popular all year long. Overnight tourists in 2018 numbered around 207,600—a relatively small number compared to the province's day tourist arrivals (Forsgård, 2019). The second most popular type of travel experience is rural and farm-based tourism (Dernoi, 1983; Johansson, 1991). The bucolic lifestyle of the archipelago and the good connections between islands promote a rural idyll that is conducive to cycling, walking, swimming, sailing, and renting vacation homes (Fig. 8.4). Rental cottages are the most popular lodging option on the islands, and renters stay significantly longer than other tourists (Mäkelä, 2009). Connected to the theme of rurality and agritourism, there is a strong sustainable food movement that has translated into culinary tourism (Fellman, 2016), and the region has a salient number of heritage sites, including churches, castles and maritime history monuments (Grydehøj, 2010).

Mount Athos

Mount Athos is a mountainous peninsula in Greece that is very rich in history and religious heritage. Although it is an integral part of the Greek state, it is treated differently from other parts of the country because of its Greek Orthodox monastic history. Its autonomy allows the monastic society there to function as a protected area for Orthodox religious practices. As part of its ascension to the European Union (the European Community at that time), Greece was granted permission to continue allowing Athos to have 'special jurisdiction' over matters of the free movement of goods and people within the territory more formally known as

Fig. 8.4. Sailing is a popular tourist activity in the Åland archipelago

the Autonomous Monastic State of the Holy Mountain (Alexopoulos and Fouseki, 2016). This was particularly important because, as a male-only monastic entity, women and children are forbidden to enter. This unusual exception to the free movement of people in the EU was granted only because the statute was seen as an important part of Europe's religious and cultural diversity and faith heritage.

Mount Athos has an area of 336 km² and a population of around 2000 men from Greece and several other Eastern Orthodox-dominated countries. The monastic 'state' lies outside the EU's VAT regime (Kovova *et al.*, 2018), largely because the Orthodox Church is a tax-free entity but also because of Mount Athos' special legal status in the EU. Even though the holy mount is part of the Schengen Area together with the rest of Greece, special permits similar to visas are required of all visitors, and it maintains strictly guarded borders with the rest of Greece.

The peninsula has been a UNESCO World Heritage Site since 1988 for its role as the Eastern Orthodox spiritual center, its autonomous status since ancient times, and for its artistic and architectural heritage (UNESCO, 2020). Tourism is an important part of the jurisdiction's economy (Kotsi, 2012; Ron and Timothy, 2019). Although entrance and monastery stays are free, many monks are engaged in handicraft and icon making, and the monies earned are usually used for the monks' sustenance and for maintaining historic buildings.

Based upon this heritage and its religious importance, the tourist market is comprised of Orthodox pilgrims, the faithful curious, and ordinary cultural tourist who are keen to see the monastic lives of the monks, the unique built religious environment and to satisfy their curiosities. The monastic state has no hotels or bars, although lodging is provided in the monasteries for up to 1200 men per night, and it has a daily quota on the number of visitors allowed, yet many of the monks feel there are too many visitors, which is beginning to crowd out the spiritual character of the place (MacFarquhar, 2018), 'destabilizing the boundaries between the sacred and the secular' (della Dora, 2012, p. 951). Approximately 300,000 tourists visit the territory each year, two-thirds of which come from Russia.

The gender-segregated nature of this self-governing monastic entity is still hotly debated in the European Union, with antagonists arguing that forced gender exclusion is un-European and against the EU's human rights mandates, causing Alexopoulos and Fouseki (2016, p. 235) to raise the question, 'should a religious site of local, national or international significance that excludes half of the world's population be designated as a heritage place of universal value?'.

Svalbard

The Svalbard archipelago is a sovereign part of Norway and one of the northernmost land areas of Europe, located in the Arctic Ocean at a latitude of 74–81°, essentially the same level as far northern Canada or northern Greenland. It has a permanent population of just less than 3000 and a total land area of 61,022 km². Sovereignty over the islands was granted to Norway in 1920, which took effect in 1925, by the Spitzbergen Treaty. As of 2020, 46 countries have signed the treaty acknowledging Norwegian sovereignty over the islands. Despite Norway's ownership of the islands, administratively Svalbard lies outside of many of Norway's jurisdictional controls. It is classified as an unincorporated part of Norway with its own governor, and many elements of its administration and daily life are governed by the Svalbard Treaty rather than the Norwegian state (Lew *et al.*, 2015).

The territory can only be used for peaceful purposes, and all signatory states (to the Svalbard Treaty) have the right to use the archipelago for economic purposes. At present, the four main economic sectors are coal mining, scientific research, fishing and tourism (Timothy, 2010). As well, citizens of any country are permitted to live and work in Svalbard without a visa and without hindrance, as long as they follow the law and have an income source. Whereas Norway is a Schengen state, the archipelago is not, and visas are not required of anyone visiting the territory. As well, passports are required on all flights to and from the islands, including from continental Norway. Svalbard has a lower income tax rate than the rest of Norway, and being outside the European Union means there is no VAT tax (Kovova *et al.*, 2018), but the region is too isolated and expensive to visit to become a retail tourism destination.

Tourism has grown to be one of the main economic foci of the archipelago. In 2018, nearly 70,000 tourists visited (Nikel, 2020), and the industry is booming. While authorities appreciate the growth in tourist arrivals over the past ten years, they are growing increasingly concerned about tourism's environmental impact and about the safety of visitors due to polar bear attacks and thawing ice. With the boom in tourism and the ease of living and working in Svalbard, several unscrupulous tour guides and outfitters have appeared. Many of these are not well trained for such remote and arctic environments, and they lack sufficient knowledge about the archipelago's ecosystem and history to make a positive impact on visitors. As such, the Norwegian government is taking actions to make it harder for random individuals to show up and start their own guiding or outfitter services (Nikel, 2020).

Much of the area's tourist appeal is its extreme location on the Arctic periphery, its dramatic mountainscapes and fjords, and its polar bear population (Timothy, 2010; Viken, 2011). As the gateway to the islands, Longyearbyen, the main population center, is the focus of tourism activities. Hiking, glacier walks, dog-sledding and kayaking are important activities in the region, and several cruise ships stop by each year as part of the recent trend toward extreme and expedition cruising (van Bets *et al.*, 2017). Svalbard also has a strong heritage component given that it has long been a base for Soviet and later Russian mining settlements (Grydehøj, 2010; Grydehøj *et al.*, 2012). This heritage has resulted in some conflicts between the two states in the past, with Russia exerting claims over some of its mining settlements. Given the open borders of Svalbard, as a precautionary move, Norway began utilizing tourism more purposefully as a tool for asserting its sovereignty over Svalbard, particularly in light of continued contestations with Russia over the use of its resources and even Russian counter-claims over territory (Timothy, 2010).

Faroe Islands

Denmark's two overseas territories, the Faroe Islands and Greenland, are autonomous territories, or autonomous 'countries'. By popular vote, Greenland adopted, and was granted, a home-rule system in 1979, which conferred a significant level of autonomy from Denmark and self-determination in areas such as education, fishing, ecology and health care. This change in governance also resulted in Greenland becoming an autonomous 'constituent country' in the Danish realm (Nielsen, 2001; Timothy and Olsen, 2001). In 2009, the home-rule law was updated, giving Greenland additional powers, including justice, business, and the possibility for future independence (Grydehøj, 2018; Hovgaard and Ackrén, 2017). Together with continental Denmark, Greenland and the Faroe Islands constitute the Kingdom of Denmark. Given its geographic location in North America, Greenland is not considered in this discussion.

The Faroe Islands' autonomy was established in the 1948 Home Rule Act. This legislation allowed the decentralization of power from Copenhagen to the archipelago's regional government to govern essentially all domestic affairs. Home rule enables the Faroese government to collect taxes and oversee its own customs services, education and health care. As the holder of sovereignty over the Faroes, Denmark remains in charge of foreign trade, defense and the high courts. Although the Faroese are Danish citizens, they do not have the same freedom of movement rights within the EU as citizens living in Denmark. The Faroe Islands are not part of the European Union or the Schengen Area (Hall, 2014). Because of its high level of self-determination, the archipelago was able to decide against EU membership owing largely to the EU's strict fishing regulations, fishing being the islands' main economic sector (Bailey, 1998; Hovgaard and Ackrén, 2017).

The 2005 Foreign Policy Act, gave the Faroe Islands a higher level of autonomy by allowing it to negotiate on an international level in certain areas not overseen by Denmark. These include taxation, financial policy, business regulation, fisheries management, natural resource management, energy use and production, environmental protection, labor markets, social security, emergency preparedness, education, scientific research, culture, customs, and external trade relations. Denmark continues to oversee constitutional issues, citizenship, supreme court, defense, and monetary and currency matters. Plans are currently being considered for the Faroe Islands to take over control of its own immigration policy (Føroya Landsstýri, 2019).

Born largely out of isolation, self-reliance, and self-determination, Faroese nationalism is strong (Olafsson, 2000). Whereas there is a strong political movement to join the EU, there is an equally strong desire among many citizens for more autonomy, including full independence. The Danish government, many Faroese politicians, and approximately half of the population supports a sovereign Faroese state. The main obstacles to a declaration of independence are psychological, a fear of the unknown, as well as losing heavy subsidies from Denmark and the EU (Thomassen, 2011, p. 82; see also Grydehøj, 2018; Hovgaard and Ackrén, 2017).

Although fishing is the main industry in the Faroes, tourism is becoming increasingly important (Prideaux, 2008). Nature is the islands' primary attraction, with windswept and treeless islands, unique wildlife and dramatic coastlines. The capital, Tórshavn, is the main gateway for air and sea arrivals and the center for accommodations and exploring the outer islands. The tourism authority, Visit Faroe Islands, categorizes the market into globetrotters, nature lovers, sightseers and cultural lovers (Plieninger *et al.*, 2018).

The unique geopolitics of the Faroe Islands affects tourism in several ways. First, the islands have their own version of the Danish currency. It is fixed to the krone's value and exchange rate, but the script on its unique banknotes is in Faroese and depicts local culture, and the monetary policy is administered by the Danish National Bank (Danmarks Nationalbank, 2014). There are ongoing discussions in the islands regarding adopting the euro and how that might happen given that Denmark controls the Faroese krone.

Second, because the Faroe Islands remain outside the EU and Schengen, the four basic freedoms of the Common Market—free movement of people, goods, services and capital—do not necessarily apply. For example, all EU citizens, except those from the Nordic States, need visas and employment permits to live and work in the archipelago. EU citizens and Nordic citizens need a national identity card or passport to enter. All others are required to have a passport, and Schengen visas (including those issued by Denmark) are not valid for entry into the Faroe Islands; Denmark issues visas specifically for the islands. As well, since the islands are outside the EU, they also have their own customs realm, and levy their own taxes, including a VAT tax, income tax and business tax.

Third, the Faroese government has ultimate authority over transportation and infrastructure policy, environmental law and cultural heritage protection. All of these areas have clear and important implications for tourism, and the protection of the country's natural scenery and traditional ways of life are among its biggest concerns when it comes to developing tourism. Because of its small size, isolation and ecological and cultural vulnerabilities, there has been considerable debate about the merits and challenges of tourism, with most of the population expressing concerns over the unorganized nature of tourism, the perception that tourism is unsafe and that tourists will disturb wildlife and nature (Plieninger *et al.*, 2018).

The British Overseas Territories and Crown Dependencies

As noted previously, several countries in Europe continue to maintain a colonial presence in the Pacific Islands and the Americas. Many of these territories are highly tourism oriented. Although a detailed description and analysis of these overseas territories is warranted, this is beyond the European continental focus of this book. Thus, only a few overseas territories and dependencies are examined here to illustrate commonalities or differences that might exist in Europe and elsewhere.

In addition to detached Northern Ireland, there are several other overseas entities in Europe that fall within the British realm. Whereas Northern Ireland is an integral part of the UK, the two British Overseas Territories (BOTs), Gibraltar and the Sovereign Base Areas in Cyprus, and the three Crown Dependencies (CDs)—Guernsey, Isle of Man and Jersey—are not part of the UK, but they are associated with it under the sovereignty of the crown. The differences between BOTs and CDs are sometimes confusing, but generally speaking, CDs have wider powers of self-governance than the BOTs (Clegg, 2018; Hendry & Dickson, 2018), nearly to the point of being independent countries, having full control of their domestic policies and a considerable degree of control over foreign policy, particularly in the area of taxes. The UK generally cannot legislate for the CDs without invitation or permission. According to the House of Commons Library (2020, n.p.), 'In contrast with the British Overseas Territories, on which the UK Parliament has unlimited power to legislate, UK primary legislation does not ordinarily apply to the Crown Dependencies'.

The history of the two types of territories also differs. All 14 of the BOTs are remnants of the British colonial empire and have chosen to remain under British rule. The CDs, on the other hand, have a longer history of self-governance. Guernsey, the Isle of Man, and Jersey are leftovers from historical times when they were self-ruled monarchies or autonomous parts of other duchies and kingdoms before the emergence of England and later the UK. Historically they had their own legislatures, elections, courts, tax regimes and legal systems, which continue independently to this day (de la Haye, 2013; King, 2013; McMahon, 2013).

Guernsey, the Isle of Man and Jersey (the CDs) are not part of the UK or the European Union. They are self-governing territories over which the UK has certain responsibilities, including military defense, citizenship, extradition laws and other internationally relevant actions. The three CDs are not represented in the British parliament but instead have their own legislative assemblies, judicial systems and laws, and fiscal systems that are entirely separate from those of the UK (House of Commons Library, 2020). Although not part of the UK, BOTs belong to the crown and have to abide by British laws, although some of them have their own adaptations to local circumstances. All of the territories have a certain level of self-governance, but the degree of autonomy varies among them. The governors assigned by the crown to represent it in the BOTs have a great deal of power to enforce UK laws overseas; the UK's rights in the CDs are more limited. An illustration of the CDs' separate authority comes from the COVID-19 pandemic of 2020. Although Channel Islanders arriving in the UK were exempt from the UK's quarantine regulations, Guernsey and Jersey have banned all tourist (non-resident) arrivals, and islanders returning from Great Britain must self-isolate for 2 weeks (TravelMole, 2020).

Although technically not part of the European Union, the European BOTs (Gibraltar, Dhekelia and Akrotiri) and the CDs (Isle of Man, Jersey and Guernsey) were *de facto* part of the EU Customs Area through their association with the UK. However, the UK's withdrawal from the EU on January 31, 2020, means that the territories are preparing for life outside the Common Market (House of Commons Library, 2020). In preparation for this event, Guernsey, Jersey, the Isle of Man, and the UK signed a customs union in 2018, which will ensure that existing four-way free trade will continue beyond the UK's departure, reaffirm the free trade area's common external tariffs, and include the CDs in any future trade agreements between the UK and the EU (States of Jersey, 2020a).

Like the CDs, Gibraltar and the Sovereign Base Areas by default have functioned in many respects as part of the EU, even though they were officially not part of the alliance or part of its customs union. Although Gibraltar participated in the EU through its association with the UK and was required to abide by EU laws and regulations, because of its small size and isolation from the UK, the territory was excluded from the common customs tariff, common agricultural policy, and VAT requirements. This gave rise to the success of gaming, duty-free shopping and offshore banking outside the rules of the EU. Given its land border with Spain, which is a full member of the EU Common Market and a Schengen state, there have always been passport and customs inspections at the border (Fig. 8.5).

Akrotiri and Dhekelia

The British Sovereign Base Areas (SBAs), Dhekelia and Akrotiri, on the island of Cyprus are a British Overseas Territory (BOT). Because of Cyprus' strategic location near the Middle East and Eastern Europe, the UK desired to maintain a military presence there while offering independence to the Cypriots. As part of the requirements for Cyprus' 1960 independence, the UK retained sovereignty over the two SBAs for military purposes (Clogg, 2015; Dodds

Fig. 8.5. Customs and passport formalities have always existed on the Spanish-Gibraltar border

et al., 2015). Dhekelia and Akrotiri comprise about three percent of the land area of Cyprus (254 km²) and have a population of around 15,500, comprising 9700 Cypriots and 5800 British service members, contract personnel and their dependents (Central Intelligence Agency, 2020). Spatio-demographically, the SBAs are home to villages and farmlands where Cypriots live and military garrisons and cantonments where British service members live. Although the boundaries of the SBAs clearly mark the lines of UK sovereignty, the Cypriots who live on SBA territory are Cypriot citizens, and Cypriot civic laws generally apply.

The SBAs have a unique government. They are administered by the Ministry of Defense rather than a civilian elected government such as in other British OTs (Dodds *et al.*, 2015). There are no local elections. The SBA government is comprised of several departments, including Law and Justice (police, courts, and attorney general), Customs and Immigration, Environment, Area Offices, and Media and Communications. The area offices, one in Dhekelia and one in Akrotiri, oversee births and deaths, planning permits, business establishments and licenses, water management, hunting permits, alcohol sales and sundry other civil responsibilities (Sovereign Base Areas Administration, 2020).

Cypriots living in the SBAs are not British citizens and must be approved to live in the area. Most of the laws of the SBAs are synchronized with the laws of the Republic of Cyprus. SBA Customs and Immigration oversees external border controls (with the occupied northern part of the island—the Turkish Republic of Northern Cyprus—and the UN buffer zone) (Fig. 8.6), taxation, asylum claims, fisheries and coastal area management, and customs service on flights arriving into the bases from the UK and third countries (Sovereign Base Areas Administration, 2020). Postal services for British citizens at the garrisons is provided by the British Forces Post Office, while in the civilian

Fig. 8.6. British Sovereign Base Area Customs and Immigration checkpoint at the border with the Turkish Republic of Northern Cyprus

Fig. 8.7. The heavily fortified border of a British garrison in Dhekelia SBA

areas of the territory, it is provided by the Cyprus Postal Service, and the local currency is the euro (Sovereign Base Areas Administration, 2020). Although the SBAs were not part of the EU, they followed most EU policies and laws, especially those pertaining to open boundaries, agriculture, fishing, and customs and fiscal policies. With the UK's January 2020 departure from the EU, it remains to be seen how the relationship between the SBAs and Cyprus will change. According to most experts, it is very likely that the UK and Cyprus will keep the status quo so that everyday life remains unchanged.

As far as tourism in concerned, the SBA administration does not promote the industry in Dhekelia and Akrotiri, and there are significant restrictions on business licenses within the SBAs. The Cyprus Tourist Organization does promote attractions and resources on the SBAs. From a tourism industry perspective, the SBA territories function as though they were part of Cyprus. There are no border checkpoints between the Republic of Cyprus and the SBAs, although there are high-security gateways at the entrances to the garrisons and cantonments (Fig. 8.7).

The SBAs are home to several well-used beaches and a handful of protected natural areas within their territory, including wetlands, salt marshes and wildlife migration zones (Dodds *et al.*, 2015) (Figs 8.8 and 8.9). The SBA Environmental Department administers these natural areas, provides interpretive media and encourages local and tourist visits. The sovereign bases are also home to several historic monasteries and archaeological sites. The SBAs adhere to Cypriot antiquities and heritage protection laws. The famous archaeological site of Kourion lies inside the UK's borders, but the site is managed as a significant tourist attraction by the Cyprus Department of Antiquities (Fig. 8.10).

Gibraltar

Clinging to the coast of Spain by a 1.2 km border, Gibraltar is a popular tourist destination in the Mediterranean with a population of 32,200 and an area of 6.7 km². In 1704, the 'Rock of Gibraltar' was occupied by British forces, and in 1713 sovereignty was granted to Great Britain

Fig. 8.8. Public beach in the Akrotiri SBA

Fig. 8.9. Salt marsh conservation area in the Akrotiri SBA

Fig. 8.10. Kourion Archaeological Site—although part of the Akrotiri SBA, it is managed by the Cyprus Department of Antiquities

in the Treaty of Utrecht (Seekings, 1993). As a British Overseas Territory, the colony exercises a high level of self-government in internal matters and its 2006 revised constitution ensures Gibraltarians the right to self-determination. According to Mut Bosque (2017), the territory functions almost like a sovereign microstate in internal affairs and appears to be in a transitional period between an Overseas Territory and an independent state. In recent history there have been referenda on the geopolitical status of Gibraltar, regarding whether or not the territory should become a sovereign state or join Spain. Spain has also proposed a situation of joint sovereignty (a condominium) with Gibraltar being shared equally between the British and the Spanish, but this idea has been rejected by the UK and the Gibraltarians (Clegg, 2016; Hall, 2020). After the UK's 2016 Brexit vote to leave the EU, which Gibraltar overwhelmingly voted against, Spain re-introduced recommendations for joint rule or titular control over the territory so that it could remain in the EU. For the most part, the votes in each plebiscite have favored keeping the status quo.

Nevertheless, Spain continues to maintain its position that Gibraltar is part of its sovereign territory, arguing that its granting ownership to Britain in the Treaty of Utrecht was invalid (Gold, 2005). Additional contention arose with Spain in 2019, as the Gibraltar government approved a land reclamation project that would provide building space for several high-rises and more than 600 apartments to add to the territory's living spaces. Spain argues that the reclaimed land would infringe on its sovereign territorial waters and has threatened to take Gibraltar to court over its decision to expand. Spain continues to assert control over the Rock. Recently (February 2019), Spanish warships threatened commercial ships to leave Spain's territorial waters as they approached Gibraltar's harbor. The provocative relationship that Spain has perpetuated over the years has affected

Fig. 8.11. Owing to the adverse relationship between Spain and Gibraltar/UK, they share a heavily fortified border

bilateral relations between Spain and the UK and hindered the development of tourism and other sectors in Gibraltar.

Tourism has long been an important part of Gibraltar's economy. However, one of the main impediments to its development for many years was the closure of the border with Spain (Fig. 8.11). In retaliation for a 1967 referendum on the territory's sovereignty in which 99.2% of the population chose to remain a BOT and the drafting of a new constitution, Spain unilaterally sealed the border from 1969 until 1982, when it opened for pedestrian traffic only. In 1985, the border fully opened to pedestrians and motor vehicle traffic (Seekings, 1993). For 13 years, the only way of entering or exiting Gibraltar was by air or by ship. During that time, Spain also imposed flight path restrictions, effectively creating an economic blockade until the issue was resolved in 1985 (Gold, 2005; O'Reilly, 1999).

Gibraltar is a thriving tourist destination, especially among British tourists, but it is particularly popular among the Spanish market for duty-free shopping and gambling (Fig. 8.12). Its marina, one of the best known in the Mediterranean, is a home port for many luxury yachts; it is also a popular cruise stopover. Other than shopping and gaming, the main appeal of the territory is the famous Rock of Gibraltar, the monolithic limestone landmass that comprises almost half of the outlier's land area (Fig. 8.13). The Barbary apes (Barbary macaques) inhabit the Rock and are considered by many to be Gibraltar's main tourist attraction (Fig. 8.14). They comprise Europe's only wild monkey population and have considerable heritage value among residents and tourists, although there is considerable concern among wildlife conservationists that the macaques population has become too friendly with tourists and too reliant upon their handouts. In 2018, some 10.4 million visitors crossed the land border and 217,059 tourists arrived by air.

Fig. 8.12. One of Gibraltar's busy duty-free shopping zones

Fig. 8.13. The iconic Rock of Gibraltar is the territory's main geographical feature

Fig. 8.14. The Barbary macaques are one of Gibraltar's most popular attractions

The British Crown Dependencies: Guernsey, Isle of Man and Jersey

There are three British crown dependencies, all located in Europe (Table 8.3). Their considerable autonomy within the British realm enables them to develop their own tourism policies, and as members of the Common Travel Area (CTA), there is absolute freedom to travel between the three island groups, the UK, and Ireland. The CTA functions entirely outside the auspices of the EU and Schengen. It began in 1922 after Ireland seceded from the UK, with the realization of the difficulties the two countries would face in controlling their common boundary. In the years that followed, legislation was enacted at various times to clarify specific terms of the CTA. When Ireland joined the EEC in 1973, and when it came time to ratify the Schengen Agreement, Ireland opted out because of its open border with the UK, which had also decided to opt out of Schengen. The CTA also allows British citizens the right of abode in Ireland and vice versa, but living and working in the CDs, even for British citizens, sometimes requires a work permit.

The Isle of Man, Guernsey, and Jersey have traditionally maintained open borders with the UK. The CDs have a common visa and customs policy with the UK. All arrivals from the UK or Ireland are treated as domestic, but anyone arriving to the Channel Islands from outside the Common Travel Area must have a passport (or identity card for EU and EEA citizens), and if a visa is necessary for the UK, it is also necessary for the CDs (States of Jersey, 2020a). With the full withdrawal of the UK from the EU on December 31, 2020, it is likely that all direct arrivals to the CDs from outside the UK will need to be in possession of a passport.

Their individual quasi-independence also empowers them to produce their own versions of the pound sterling as part of their autonomous monetary policy, and operate their own postal services. The only things the islands do not govern for themselves are citizenship, military security, and international agreements.

Tourism in the Isle of Man and Channel Islands is rather unique. Because they were located outside the tax and fiscal regimes of the European Union with no VAT requirements, all three territories developed thriving offshore banking industries (Canavan, 2014, 2017; Entwistle and Oliver, 2015; Guiberteau, 2019; Hampton and Christensen, 2007; Oliver, 2019), although this has come under increased scrutiny in recent years as concerns have arisen over the increased potential for money laundering and tax evasion in the three areas (Entwistle and Oliver, 2015).

Table 8.3. General information about the British Crown Dependencies, 2020

Territory	Population	Size	Location	Capital city
Guernsey	67,052	78 km²	English Channel	St Peter Port
Jersey	101,073	116 km²	English Channel	Saint Helier
Isle of Man	90,499	572 km²	Irish Sea	Douglas

Source: Central Intelligence Agency (2020)

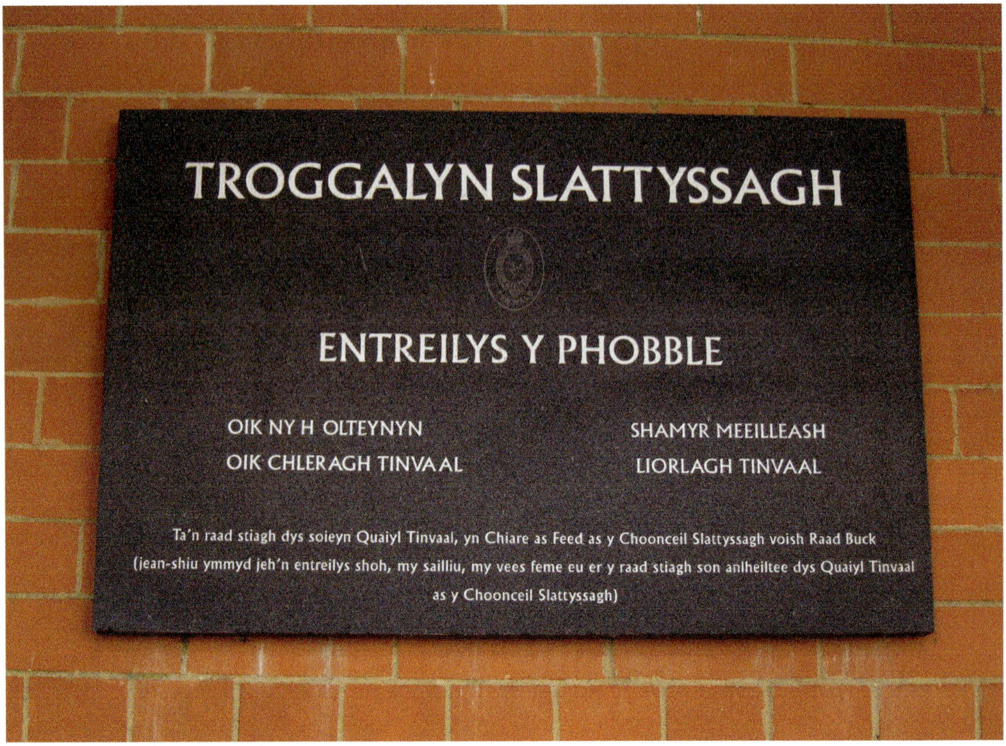

Fig. 8.15. The nearly extinct Manx language is being used more as a means of protecting the Isle of Man's cultural heritage

The Isle of Man is located in the Irish Sea between Great Britain and Ireland. It has a population of 83,300 and an area of 572 km² and had its own indigenous language, Manx, until the last native speaker died in 1974. Since that time, as part of a movement and effort to protect the Manx identity, there has been an active effort to resurrect Manx, a Gaelic language, and lessons in at least one school are taught exclusively in Manx. Despite efforts to teach the language in schools, broadcast the language on the radio, and produce bilingual books and signs throughout the Isle of Man, only around two percent of the island's population has a functioning knowledge of the language (Fig. 8.15).

The Isle of Man (sometimes referred to as Mann) and its Gaelic population have been governed by many rulers and settled by various people over the past millennium, including the Norse, the Scots and the English. There is a significant built heritage related to these successive rulers (Fig. 8.16), but predating those sites are remnants of Neolithic civilizations scattered throughout the island which, together with the Manx Museum, are the main cultural heritage attractions (Figs 8.17 and 8.18). The verdant rural landscapes and villages with agritourism and farmstays, local history and folklore, as well as natural scenery form much of the island's tourism appeal (Fig. 8.19). This was enhanced when the island was designated a UNESCO Biosphere Reserve in 2016 for its remarkable coastal environment.

Tourism in Mann boomed during the late 1800s and early 1900s with the industrialization of the British Isles. The island became known as an affordable coastal retreat for the working class of the industrial cities of Great Britain. It reached its peak in 1913 when over 600,000 people visited the island, but with the arrival of inexpensive packaged trips to the Mediterranean, the popularity of Mann began to decline (Smith, 2019). This, according to Canavan (2013, 2015a, 2015b), has heightened the Manx sense

Fig. 8.16. Peel Castle is one of the Isle of Man's main heritage attractions

Fig. 8.17. Remnants of Neolithic life dot the landscape of Mann

Fig. 8.18. The Manx Museum tells the story of what makes the Isle of Man unique

Fig. 8.19. The countryside of Mann provides a rural idyll that appeals to most visitors

of marginality and undermined local identity. Since the late twentieth century, the island's tourism authorities have worked to promote the island's old-time atmosphere, its natural appeal and its history. Unlike in most small territories, where domestic tourism is either negligible or uncounted (Timothy, 2020), Canavan (2012) identifies domestic tourism on Mann, which he terms 'microdomestic tourism' because of the island's small size, as a salient form of local tourism. Microdomestic tourism entails day trips and overnight stays that are touristic in nature and a phenomenon authorities should consider in policy-making to help ensure the maintenance of facilities, infrastructure and attractions to make the island more attractive to foreign and domestic visitors. Timothy (2020) argues that domestic tourism in microstates and small territories is a different phenomenon entirely than it is in geographically larger entities. Canavan (2012) provides empirical evidence to support this view of domestic tourism on the Isle of Man.

In common with the Isle of Man, the Channel Islands are highly dependent on the British domestic tourism market (Cooper, 1995). Their historical separateness, unique cultures, tax benefits and insular environments are what attract tourists to the Bailiwick of Jersey and the Bailiwick of Guernsey (which also includes the populated islands of Alderney, Herm and Sark). Historically, the two territories were culturally and politically connected to Normandy in today's France. In the early thirteenth century, when King John of England lost his mainland Norman territories to the French King Phillip II, the Channel Islands remained possessions of the English king.

Unique in the British Isles, the current legal systems in the two Channel Island CDs are founded on medieval Norman law, combined with elements of contemporary English common law and French civil law (de la Haye, 2013; McMahon, 2013). Jersey Islanders and Guernsey Islanders originally spoke their own versions of the Norman language, a Romance language closely related to French. Jèrriais and Guernésiais are still spoken by a small minority of the islands' populations, but these local dialects have been gradually replaced by English (Wilson *et al.*, 2015). English and French (Jèrriais) are the official languages of Jersey, and English and French (Guernésiais) are the official languages of Guernsey. Numerous street signs are written bilingually, and many legal documents are still written in French.

Guernsey's economy is geared more to agriculture and light industry than Jersey's economy, which is geared more toward offshore banking and finance. Tourism is a crucial part of both entities' economies and melds well together with farming and banking (i.e. rural and agritourism, and business travel). With no VAT tax levied in the Channel Islands, certain luxury items have traditionally been cheaper than in the UK and France, which stimulated a fairly robust retail tourism sector as well. Other taxes, however, have been levied by the bailiwick governments over the past 20 years, which have increased island prices and reduced their competitive retail advantages.

Guernsey and Jersey saw their tourism zenith in the 1980s and 1990s, particularly the British market, although they have seen some decline, and both bailiwicks have made strides to rejuvenate their tourism sectors (Chaperon, 2017). A large part of Jersey's popularity as a destination in the 1980s and 1990s was due to it being the location of the hit British television show, *Bergerac*, which was filmed on the island from 1981 to 1991. Today, fans of the show continue to visit Jersey as a manifestation of film-induced tourism (Jersey Heritage, 2020). Closely related to film-induced tourism is literary tourism, which has been a boon for Guernsey, where Victor Hugo lived for 15 years and wrote several famous works, including *Les Misérables* and *Toilers of the Sea* (Leadbeater, 2019). In 2017, Guernsey received 1,134,568 arrivals by sea and air (although this included domestic arrivals from Alderney) (States of Guernsey, 2018). In 2017, 726,800 tourists visited Jersey (States of Jersey, 2020b). Both dependencies are thriving tourist destinations with well-organized destination management organizations (DMOs), a wide range of attractions and activities, and many lodging and eating options.

Conclusion

Autonomous regions have a unique relationship with the state. They are self-governing over internal matters, with the state-level entity only acting on their behalf in terms of citizenship, high courts, international diplomacy, currency controls, and military defense. Many autonomous regions in Europe are continuing to receive a deeper level of independence, and there are some efforts towards secessionism. This, however, would mean losing the protection and fiscal support of the mother country, so most secessionist movements are highly symbolic as a means of garnering additional devolved powers at the local level.

The British crown dependencies also have considerable autonomy within the realm of the Crown. Because of their high level of internal governance, they are less beholden to the Crown than the overseas territories are, and they have chosen to focus on tourism as an important development tool, including offshore banking and finance. It remains to be seen how the UK's January 2020 exit from the European Union will affect their autonomy and ambitions for more. Many of the British overseas territories have elected to remain possessions of the UK, largely because of the financial incentives provided by the EU. The decision to leave might cause some of them to rethink their relationships with Britain. It is also unclear at this point how Brexit will affect tourism to Gibraltar, the Channel Islands, the Isle of Man and the Cyprus SBAs.

9

International Exclaves and Extraterritorial Places

Introduction

There are many small and unique geographical oddities that defy traditional conceptualizations of territory and the state. The very existence of overseas territories described in Chapter 8 and microstates in Chapters 3 to 7 attests to this fact. However, there is another geopolitical phenomenon that is the subject of international law and spatial analysis that at first defies all logic: exclaves or enclaves. Most people have not heard of exclaves, and the word 'enclave' in tourism studies is usually used to refer to areas of high tourism intensity, such as resorts and other densely developed tourism areas, which often have prescribed and bounded territories that separate tourist activities and spaces from those of the everyday lives of local residents. The word is also sometimes used (along with ethnic islands) in the social sciences to denote areas where distinct cultures or ethnicities cluster apart from the dominant society that surrounds them, often in the context of migrant groups in settler societies.

This discussion, however, focuses on the geopolitical and legal phenomenon of enclaves and exclaves, which are small pieces of a country that are physically detached from the state to which they belong and which can be accessed only by traveling through a foreign country. A more in-depth definition is provided below, as are specific types of exclaves and their unique geo-spatial and social characteristics. Almost all of the exclaves/enclaves of Western Europe have very small economies and are largely dependent upon tourism. This chapter highlights these unique geographical oddities in Europe and their socio-economic, political and physical characteristics that make them unusual and which make them of interest to tourists and other cross-border travellers.

International Exclaves/Enclaves

Perhaps the most unique and intriguing territorial anomalies in Europe are exclaves and enclaves. There are several different types of exclaves, and attempts have been made since the 1950s to create enclave typologies, including the work of Raton (1958), Robinson (1959) and Vinokurov (2007). Each of these assessments is of considerable value, and a mix of their perspectives forms the foundations of this discussion.

Exclaves are pieces of a country's territory that are detached, embedded inside another country or disconnected from the homeland by another state or more than one other state

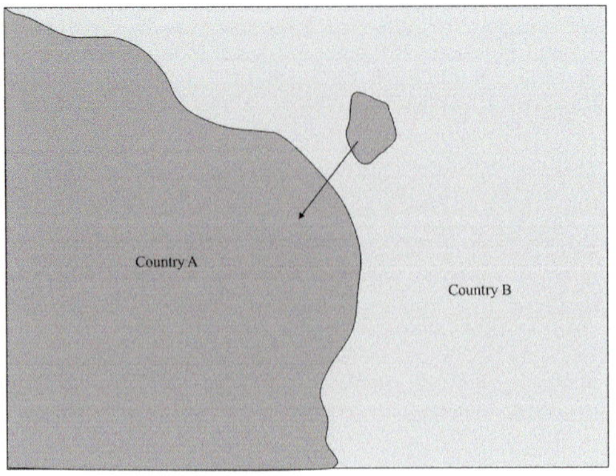

Fig. 9.1. The concept of an enclave/exclave in political geography

(Glassner and Fahrer, 2004; Lundén, 2012). From the perspective of the country to which the outlier belongs, it is an 'exclave', with the prefix 'ex' indicating a position outside or away from. Enclaves are also exclaves and are defined as fragments of one sovereign state that are physically detached from their homeland and surrounded entirely by the territory of single neighbouring state (Robinson, 1959) (Fig. 9.1). From the standpoint of the country that surrounds the outlier, it is an 'enclave', with the prefix 'en' indicating being within. Thus, both terms refer to the same object but reflect the perspective of location and direction of reference. Although it is possible for an exclave to be bordered by more than one foreign state, the same is not true for an enclave. However, in this chapter, the two terms are used interchangeably.

There are other geographical situations that are very similar to true (or normal) exclaves but deviate in certain ways (see Table 9.1). *Pene-exclaves* (or semi-exclaves) are portions of a country that are physically attached but topographically isolated (e.g. because of mountains, swamps or water bodies) so that easy access, especially by wheeled traffic, can only be achieved by passing through the territory of a foreign country (Robinson, 1959) (Fig. 9.2). *Quasi-exclaves* are technically true enclaves disconnected from the homeland, but they are so functionally and almost spatially connected that they do not function as exclaves (Glassner and Fahrer, 2004). The main example of this type in Europe are five small German territories separated from the rest of Germany only by the width of a Belgian railway line. *Temporary exclaves* result from ceasefire agreements and territorial occupations, with the understanding that they are not permanent. Until 1990, West Berlin had several small exclaves within East Berlin, which were accessible by heavily guarded corridors. *Counter enclaves* are true enclaves, but are even more unique in that they are enclaved within other enclaves (Fig. 9.3). Examples of these exist in the Middle East (e.g. Madha, UAE, and Nahwa, Oman) and Europe (e.g. Baarle Hertog/Baarle Nassau). Former or *relict enclaves* are those that were once exclaves but have now been integrated into the enclaving state (e.g. Verenahof (was Germany, now Switzerland) and the Cooch Behar area (India-Bangladesh)) or which were once pene-exclaves but now have direct road access to the homeland. Finally, *coastal exclaves* (nonsovereign semi-enclaves in Vinokurov's (2007) classification) are coastal territories separated from their homeland by a neighbouring state or international waters (e.g. Alaska (USA), Cabinda (Angola) and Kaliningrad (Russia)) (Fig. 9.4). This indicates that access is less problematic, because in most instances, maritime traffic provides approach between motherland and exclave through international waters or by agreement through the national waters of a neighbour. Independent countries, such as Lesotho, San Marino and Vatican City, which are completely surrounded by a single state, are not enclaves or exclaves but rather *enclaved states*, because they are sovereign states, not part of any other national entity.

There have been hundreds, perhaps thousands, of international enclaves throughout history, some of which have been dismantled as recently as 2015. That year a massive territorial land exchange between India and Bangladesh took place. This diplomatic marvel smoothed an extremely complex system of nearly 200 exclaves, counter exclaves, and counter-counter exclaves near their common border, some of which were no larger than a small rice paddy. This geographical labyrinth had plagued the region for decades with questions of sovereignty, resource rights, immigration, law enforcement, taxes and citizenship. Other exclaves have been absorbed by enclaving states or exchanged for similar territories across the border. Several pene-exclaves have ceased to be pene-exclaves as access has been granted by road and tunnel

Table 9.1. Example of Enclaves and Exclaves

Name of outlier	Type of outlier	Sovereign state	Enclaving state	Tourism orientation
Baarle Hertog	True enclave	Belgium	Netherlands	Yes
Büsingen	True enclave	Germany	Switzerland	Yes
Campione d'Italia	True enclave	Italy	Switzerland	Yes
Llívia	True enclave	Spain	France	Yes
Madha	True enclave	Oman	United Arab Emirates	No
Artsvashen	True enclave	Armenia	Azerbaijan	No
Međurečje	True enclave	Bosnia Herzegovina	Serbia	No
Xylotymbou	True enclave	Cyprus	UK-Sovereign Base Area	No
Sokh	True enclave	Uzbekistan	Kyrgyzstan	No
Baarle Nassau	Counter enclave & true enclave	Netherlands	Belgium	Yes
Nahwa	Counter enclave	UAE	Oman	No
Ruitzhof	Quasi enclave/true enclave	Germany	Belgium	No
Northwest Angle	Pene-exclave	USA	Canada	Yes
Point Roberts	Pene-exclave	USA	Canada	Yes
Jungholz	Pene-exclave	Austria	Germany	Yes
Kleinwalsertal	Pene-exclave	Austria	Germany	Yes
Os de Civís	Pene-exclave	Spain	Andorra	Yes
Hinterriß	Pene-exclave	Austria	Germany	Yes
Drummully salient	Pene-exclave	Ireland	UK	No
Verenahof	Relict enclave	Germany	Switzerland	No
Samnaun	Relict pene-enclave	Switzerland	Austria	Yes
Kaliningrad	Coastal exclave	Russia	Poland and Lithuania	Yes
Ocussi	Coastal exclave	Timor Leste	Indonesia	No
Teburong District	Coastal exclave	Brunei	Malaysia	Yes

construction, and in some areas, the opening of borders has rendered some exclaves less relevant than in the past. Today, dozens of exclaves, pene-exclaves, quasi-exclaves, relict exclaves, and coastal exclaves exist in nearly every corner of the globe, with important political, social and economic implications, including tourism. This chapter does not discuss coastal exclaves or enclaved states.

Exclaves have come to exist through various geopolitical processes, such as state dissolutions (e.g. USSR and Yugoslavia), the decolonization of colonial realms, mistakes in surveying boundaries, or through military aggression (Borthakur, 2017; Castan Pinos and Jagetić Andersen, 2012; Kwiatkiewicz, 2012; Mkrtchyan, 2012; Osiewicz, 2012). However, most of Europe's exclaves, at least the ones of

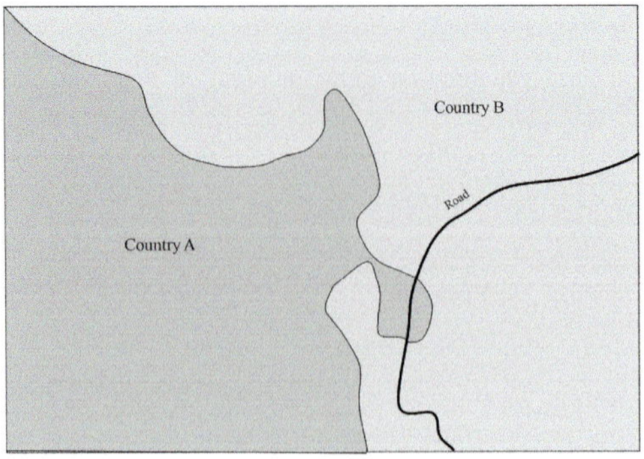

Fig. 9.2. The concept of a pene-enclave/exclave in political geography

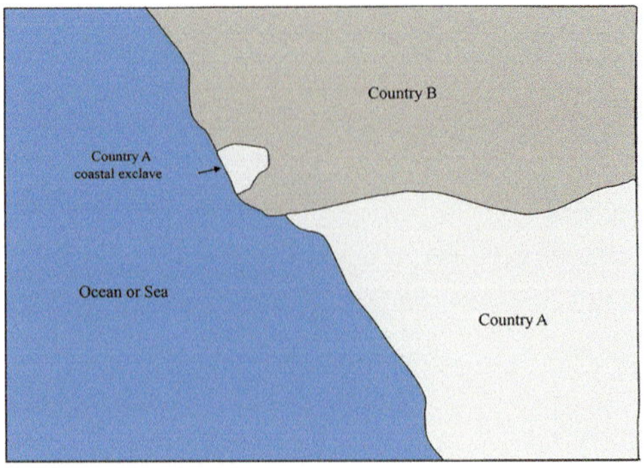

Fig. 9.4. The concept of a coastal exclave in political geography

focus in this chapter, evolved through changes from feudalism into bounded states from 1500 to the mid-1800s, 'when land ownership and fiefdoms eventually turned into politically defined states' (Lundén, 2012, p. 12). During medieval times, the concept of borders was somewhat different from what it is today. Under the feudal system, governance was based on administrating a network of obligatory relationships between nobility (lords), vassals (someone who owed loyalty and service to the landlord), and fiefs (inheritable lands held by the vassals in return for their services to the lord). Concepts of territoriality were based on social class, land ownership, loyalty and vassalage, with a lot of land in Europe being transferred from noblemen to vassals as gifts and rewards in reciprocal relationships (Reynolds, 1996; Stephenson, 1942).

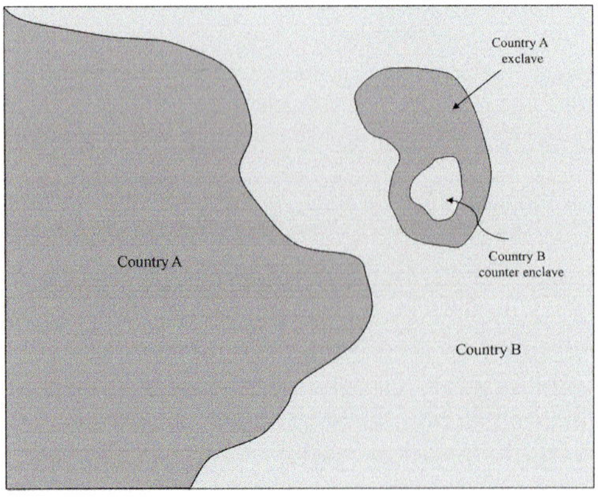

Fig. 9.3. The concept of a counter-exclave in political geography

Later, lands were exchanged between landholders, and rulers of empires frequently purchased properties from other rulers. As states and their boundaries became more clearly defined, the landholdings of monarchs and vassals formed the modern notion of statehood.

Separateness, enclavement and smallness are the common characteristics of true enclaves (Forysinski, 2012). Most exclaves are very small in area, the majority being less than a few square kilometres (or hectares), and also small in population, frequently with less than 1000 inhabitants. They are typically comprised of a single village and some rural hinterland, relatively close to their home countries, separated by only a few hundred meters in many cases. The majority of European enclaves, including pene-enclaves, are rural and have depended historically on agriculture as their primary livelihoods. Most European exclaves receive most utility services from the enclaving state, although administrative responsibilities have customarily been a mixed effort between host and home state. Law enforcement typically comes from the home country, with bilateral provisions that allow foreign police to traverse the territory of the host state. The majority of them have a dual currency system, with the currency of the host state being the primary legal tender, and their economies (including VAT taxes and duties) have been more closely connected to the enclaving state than to the state to which they belong. The legal conditions of exclaves are usually coordinated in one of two ways: through bilateral agreements or through customary norms (Forysinski, 2012). The most pressing historical problem and legal issue for exclaves has been the right of passage through surrounding foreign territory, so that the enclave is not held 'hostage to current relations between the two states in question' (Forysinski, 2012, p. 27; see also Catudal, 1979; Vinokurov, 2007).

Exclaves have received considerable academic attention in the political geography and law literature but less so in the tourism literature, likely owing to their small size and unassuming tourism positioning compared to the millions of other destinations worldwide. There are international enclaves in every corner of the globe (Borthakur, 2017; Catudal, 1974, 1975; Vinokurov, 2007). Worldwide, very few have tourism-based economies, whereas in Europe, North America and Africa most of them are overwhelmingly dependent on tourism (Berger, 2010; Bjelland, 2016; Gelbman and Timothy, 2010, 2019; Timothy, 1996, 1998b, 2000), and several others have significant potential to become tourist destinations (Timothy, 2017f).

The fundamental attraction common to Europe's exclaves and pene-exclaves are their unusual legal and political status, and the heritage and history that created them. This is a major curiosity factor for people who learn about them (Berger, 2010; Timothy, 2000). However, each one of them has its own personality, legal and policy constraints, physical and cultural features, and desire to promote tourism, which differentiates them from one another. Figure 9.5 illustrates seven ways in which exclaves emanate touristic appeal: a unique history and heritage, physical border markings, socio-cultural and economic differences, legal and regulatory differences, municipal and planning differences, the natural and cultural landscapes of the area, and tales and folklore associated with life in the enclave. These elements will come into play as the exclaves are described in the sections that follow.

Enclaves are more popular destinations for host country residents than for home country compatriots, because they offer a degree of 'intrigue associated with crossing an international boundary into a different cultural environment and economic realm but without having to experience border-crossing formalities' (Gelbman and Timothy, 2011, p. 115). The implementation of the Schengen Agreement and the European Common Market has dampened this allure in some respects but in other ways, the exclaves remain competitive destinations. Although these things, especially legal issues, create spaces that are conducive to tourism, enclaves face several challenges in their desires for tourism, including an imbalance of power between home and host state, small size constrains physical growth, isolation from the domestic market and stiffer

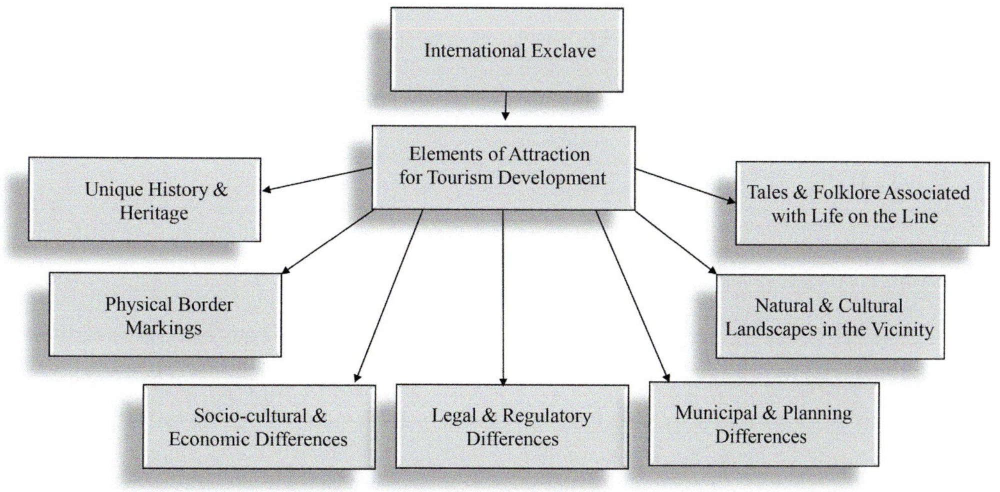

Fig. 9.5. Elements of tourism attractiveness in international exclaves (Source: partially based on Gelbman & Timothy (2011))

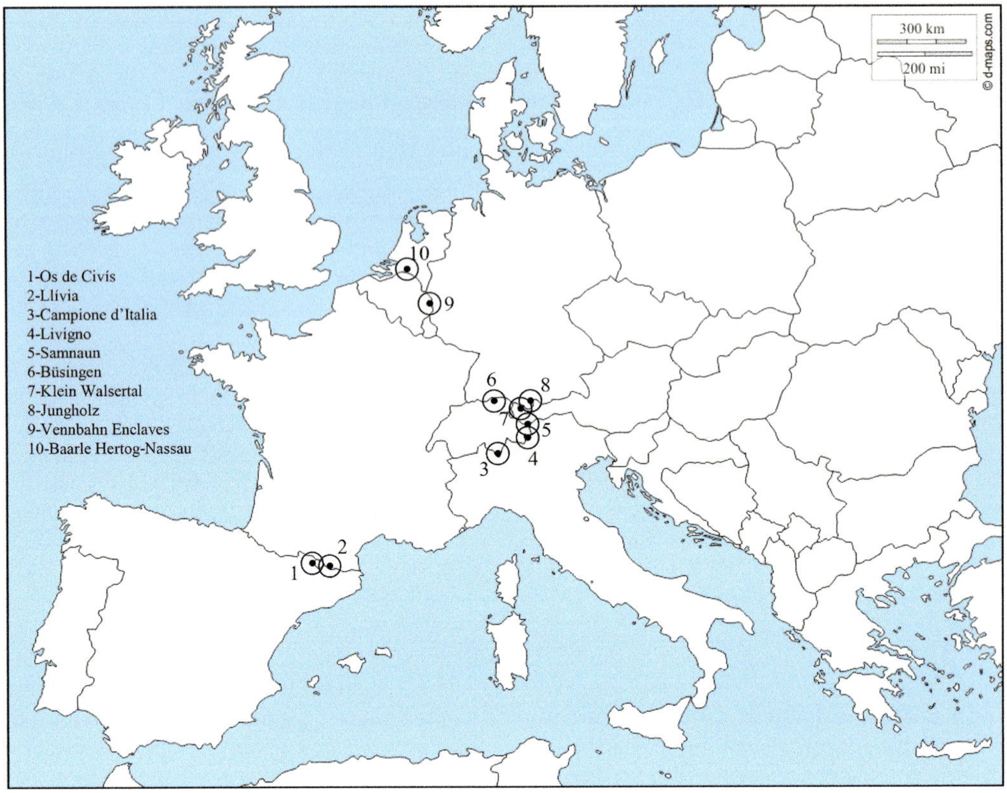

Fig. 9.6. Locations of European exclaves examined in this book (Source: basemap from d-maps.com)

competition with larger destinations that might have a more competitive geographical advantage (Timothy, 1995, 1996). The sections below examine several of Europe's tourism-oriented exclaves (Fig. 9.6).

Baarle-Hertog and Baarle-Nassau

The twin 'Baarles' are true exclaves and enclaves in every respect. This place is so unique that it has been featured many times in the popular travel media as one of the most intriguing (and quirkiest) small places to visit in Europe (see Eames, 2017; Gardner and Kries, 2005; Patawary, 2012). Baarle-Hertog is a Belgian municipality in Antwerp Province, which borders the Dutch province of North Brabant. Baarle-Hertog is comprised of four areas in Belgium proper and 22 exclaves surrounded by the municipality of Baarle-Nassau in the Netherlands, which itself is comprised of its main section in the Netherlands, one outlier entirely surrounded by Belgium proper and seven counter-enclaves wrapped within the Belgian exclaves of Baarle-Hertog (Fig. 9.7). Baarle-Hertog is 748 hectares in area with some 2250 residents. Baarle-Nassau is 7461 hectares and has a population of approximately 6500. The village of Baarle-Hertog/Baarle-Nassau lies 3.25 km from the main Dutch-Belgian border (Fig. 9.8).

The complex labyrinth of Baarle is a result of multiple medieval treaties, land trades and property sales between the Dukes of Brabant and the Lords of Breda. In the twelfth century, the Duke of Brabant transferred certain small tracts of farmland to the Lord of Breda as fiefs, but he decided to retain larger properties in the same region. Based on this decision, these small and scattered land holdings became part of Breda, surrounded by areas of Brabant. The situation remained during the periods of

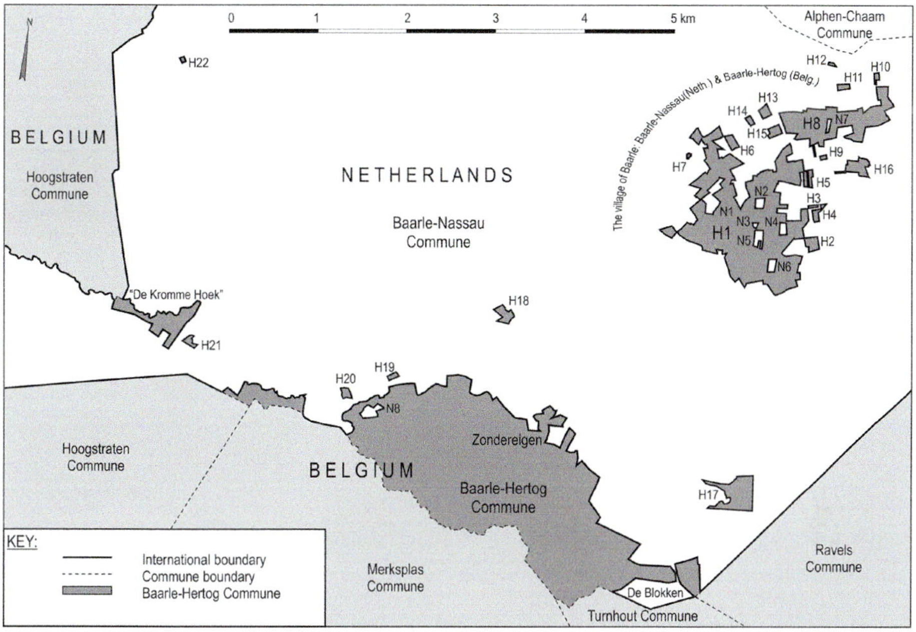

Fig. 9.7. The complex labyrinth of exclaves and counter exclaves in Baarle Hertog-Baarle Nassau (Source: courtesy of Brendan Whyte)

the Austrian and Spanish Netherlands and during the Napoleonic occupation. Following French rule, Baarle was still divided by provincial boundaries within the Netherlands, and with Belgium's 1830 independence from the Netherlands, the provincial borders were used as the international borders, and the area was divided between the two countries (Catudal, 1979; Whyte, 2002, 2004). The Maastricht Treaty of 1843 determined the international border, largely as it had functioned provincially for centuries, but the Baarle puzzle could not be resolved, so the two governments decided to allocate nationality based on each property parcel. When the Dutch-Belgian border was demarcated in the 1840s, the Baarle area was so confusing, it was left out of the process (Gelbman and Timothy, 2011; Geluwe, 2003). The enclavic border was officially defined and demarcated as recently as 1995.

The community's bewildering patchwork of enclaves and counter-enclaves has some very unique implications for everyday life and for tourism. Some of the Belgian exclaves and Dutch counter-exclaves are extremely small, encompassing only a single paddock, garden or building. Several homes and businesses are divided by the enclavic borders, which has historically created advantages and disadvantages for residents and curiosities for tourists (Gelbman and Timothy, 2011; Timothy and Gelbman, 2015). To understand the detail and seriousness with which the national border labyrinth in this community is taken, in March 2020, the Belgian government ordered the closure of all non-essential stores, but the Dutch government did not. As a result, the town's non-essential retail businesses that are bisected by the frontier cordoned off the portion of their merchandise areas that are in Belgium with caution tape, while the Dutch portions of the same shops remained open (Chini, 2020). According to one Dutch customer quoted in Chini's (2020, n.p.) report, 'I needed underwear, but that's in the Belgian part of the store, so I could not get it'. As of May 31, 2020, these orders were still in place. Border stories such as this one abound and have become an important part of the community's heritage lore. Tales of cross-border smuggling,

Fig. 9.8. A dual-nationality welcome sign at the border of the enclaved community

law enforcement challenges, and the Belgian exclaves functioning as First World War hideouts, when Belgium was occupied by Germany but the Netherlands was not, are ingrained in the local identity (Vinokurov, 2007).

As noted above, several homes and businesses are bisected by the border, and many next-door neighbours live in different countries. The address and location of divided buildings are determined by the location of the front door, and each house has either a Dutch address plate or a Belgian address plate. The front door of one historic house itself is divided by the border so that the home has two parallel addresses (Fig. 9.9). This home has become an important tourist attraction. The Baarles' heritage is in fact its border condition. In 1995, some sections of the town's international border were marked for administrative purposes and to help curious tourists find the border and trace its multiple courses through town.

The Baarles have become a popular shopping destination at the regional level (Gelbman and Timothy, 2011; Timothy, 1995). For locals, the differential regulations and laws work to their advantage. Most residents have learned to fulfil prescriptions at Belgian pharmacies and fill their car tanks at Belgian petrol stations, and purchase flowers, cheese, butter and spirits from Dutch shops (Vitaliev, 2008). These local differences translate well for tourists. Despite both countries being in the EU and Schengen Area and both having a 21% VAT rate, merchandising advantages are still notable, and certain alcohol products are taxed differently. On opposite sides of a single street in the Baarles, or in some cases in the same shop that is divided by the border, alcohol rules and tax rates differ. Although wine is taxed almost the same in both countries, beer is taxed slightly lower in Belgium. High-alcohol spirits, however, are taxed at a much higher rate in Belgium than in the Netherlands (Angus *et al.*, 2019). This adds a unique retail appeal to both sides of town, and especially for *De Biergrens* (the Beer Border) alcohol shop, which is advantageously divided by the international border (Fig. 9.10). As well, certain commercial activities are more permissive in the Dutch parts of town, such as erotica shops, while

Fig. 9.9. This divided home and front door with two difference addresses is an attraction

Fig. 9.10. This alcohol shop straddles the boundary, which offers major liquor tax advantages

Belgian chocolates are more plentiful and less expensive in the community's Belgian shops. The adoption of the euro in both territories in 2002 eliminated a significant accounting barrier (although it eroded some price advantages), standardizing most prices and helping to unify the local economy that had until that time been a mix of Belgian francs and Dutch guilders. In Jańczak's (2012, p. 67) estimation, the euro has 'unified the cities' even more.

Hours of commerce and Sunday trade were traditionally an issue between the Netherlands and Belgium. Shops in the Netherlands were required to be closed on Sundays but not in Belgium. However, owing to its unique geopolitical situation, shops in Baarle are open all weekend, and many of the opening hour restrictions for restaurants and bakeries in the Netherlands and Belgium do not apply in Baarle (Jańczak, 2012; Vinokurov, 2007). The bending of laws and traditions in this binational town has created a competitive advantage for Sunday shopping and for the types of products and services that can be purchased. The notion that 'taxes in Belgium and the Netherlands differed sometimes a lot, so one could go shopping between two tax-regimes on a single street' (Patawary, 2012, n.p.) was a major regional draw from both countries. Likewise, 'there was a time when according to Dutch laws, restaurants had to close earlier. For some restaurants on the border it meant that the clients simply had to change their tables to the Belgian side' (Patawary, 2012, n.p.) (Fig. 9.11).

Although the two municipalities have two different police forces, mayors offices, planning agencies, post offices, national banks and schools, there is a lot of collaboration between them to facilitate these and other public services. For instance, different utility services are provided for both municipalities; costs are pro-rated and divided between them depending on population or street length (Whyte, 2002, 2004). The two towns share a single fire brigade, infrastructure construction and maintenance, cultural centre and library (Municipality of Baarle-Hertog, 2020). Either Dutch or Belgian

Fig. 9.11. This popular restaurant almost straddles the international border in the Baarles

police officers can enforce laws and write tickets in any of the exclaves. Legally, Dutch police cannot carry firearms into Belgium and vice versa, but at the local level, this law is typically overlooked for expediency (Whyte, 2004).

Although the towns have separate municipal websites, from a tourism marketing perspective, both communities work together and pool their resources. They share a tourism office, publish common promotional materials, share a webpage, and work together to bring consumers to the area for the benefit of the entire town (Municipality of Baarle-Hertog, 2020).

Besides the Baarles' border web and its retail advantages, there are concerted efforts to develop other forms of tourism based on the area's rural landscapes and historic buildings. A network of cycling routes has recently been developed to explore the area's natural and rural landscapes but also to highlight the enclavic characteristics of the borderland (Timothy and Boyd, 2015). The trails weave in and out of Belgium and the Netherlands multiple times, and interpretive signage helps cyclists appreciate the unique geopolitics of this little corner of Europe.

Büsingen am Hochrhein

The tiny German village of Büsingen am Hochrhein is surrounded entirely by Switzerland, bordering the Swiss cantons of Schaffhausen, Thurgau and Zürich. The village has an approximate population of 1500 with a total land area of 7.62 km² (Gemeinde Büsingen, 2020). It lies approximately 700 meters inside Switzerland from the Swiss-German border on the Rhine River, and although it is part of Germany's sovereign territory, it functions economically as part of Switzerland (Fig. 9.12). Because of this and also owing to its location enclaved inside of Switzerland, it belongs to the Swiss customs area rather than that of the EU, and the much lower Swiss VAT (7.7%) applies rather than the higher German rate (19%) (Kovova *et al.*, 2018; Timothy, 2017b). There have not been customs and passport inspections between Switzerland and the enclave since 1947, and when Büsingers cross from Germany to Switzerland, they are treated as residents of Switzerland and are subject to Swiss customs rules. When entering Germany proper, they are equally subject to

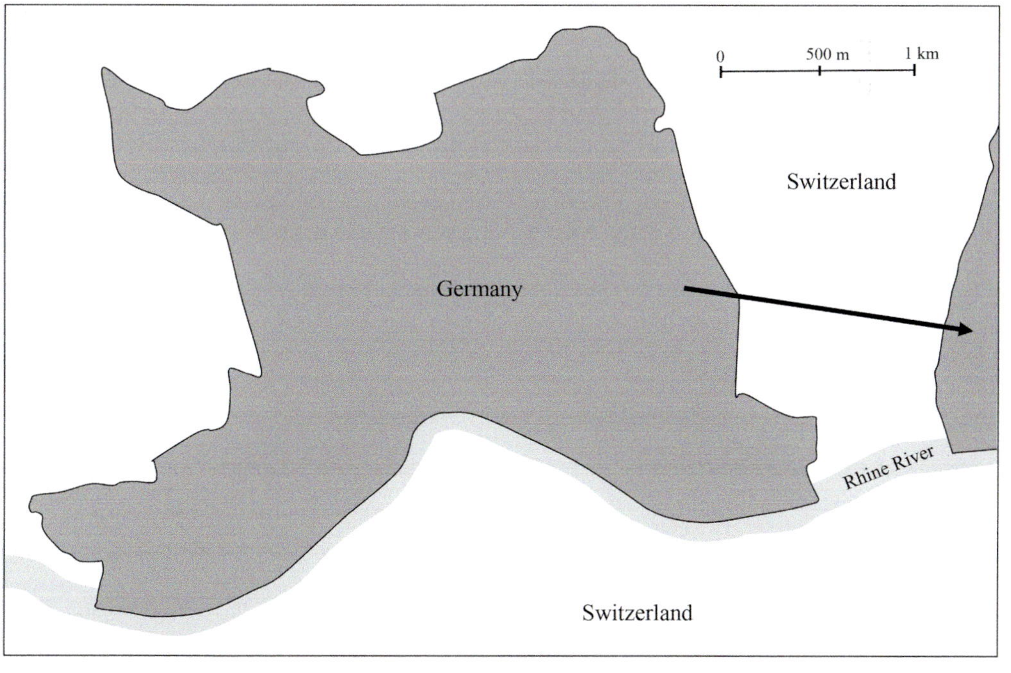

Fig. 9.12. The location of Büsingen, Germany, enclaved in Switzerland

Fig. 9.13. At Büsingen's German post office, customers can use Swiss or German postal services (photo courtesy of Hugh Wallis)

EU customs regulations (Gemeinde Büsingen, 2020).

Based on a 1964 treaty, German law applies to the exclave in nearly every respect, but Swiss economic policies and laws apply regarding customs and import duties, banking and currency regulations (Musiał-Karg, 2012). The currency of the enclave is the Swiss franc, and most employees there are paid in francs rather than euros. The village has two postal codes, a German one and a Swiss one. In the post office, Swiss and German postal services are available; residents can choose which one to use, depending on the rates for the destination of their packages (Fig. 9.13). The exclave also has two telephone area codes and its public transportation needs are serviced by Switzerland's PostBus. Büsingen has a municipal webpage, available with both German and Swiss URLs.

In 1535, King Ferdinand I of the Holy Roman Empire gifted the area of Büsingen as a fiefdom to a family of Schaffhausen as a token of gratitude for their service to the crown. This same family continued to own title of the territory until the nineteenth century (Catudal, 1979). Over the years, Schaffhausen tried several times to integrate the enclave, but the Austrian Hapsburgs and then Germany resisted as a form of castigation against Schaffhausen for its earlier abuses related to the vassal Im Thurn family. In the early nineteenth century, Büsingen came under the sovereignty of the Grand Duchy of Baden and later the German Empire in 1871 (Catudal, 1979). There have been several attempts by Canton Schaffhausen, as well as by the Büsingers themselves to become part of Switzerland (Bolli, 1954). The latest attempt in the 1960s was unsuccessful, and the final arbitration on the status and composition of the enclave was legally set by agreement between Switzerland and Germany in 1967. This treaty placed Büsingen within the Swiss customs area and eliminated border checkpoints between the enclave and its enclaving state.

Fig. 9.14. Büsingen's village center (photo courtesy of Hugh Wallis)

The enclave situation is the area's most salient tourist attraction, and its border with Switzerland is its main curiosity. The 123 historic border markers are an important heritage element and form a trail that visitors can follow along the perimeter of the enclave. The community's 'exclave trail' is mapped and signposted for visitors to learn about the territory's unique history. It focuses on the issue of enclavic sovereignty and borders, as well as the village's other historic sites, such as historic post-and-beam houses, the mountain church and the Rhine Mill (Fig. 9.14). Switzerland takes care of most tourism marketing for the enclave. In fact, the German state to which Büsingen belongs, Baden-Württemberg, makes no mention of the village in its recent promotional materials. The community has several guest houses and a hotel, with several restaurants (Gemeinde Büsingen, 2020). Of particular touristic interest is the Restaurant Waldheim, where tourists can sit at a table 'where the border runs beneath their feet' (Switzerland Tourism, 2020). Tourism is currently a top development focus of Büsingen's leadership. The local council aims to develop 'soft tourism' based on the 'idyllic landscape' of the area on the Rhine and the village's 'special status as an exclave' with closer linkages with the rest of Germany (Gemeinde Büsingen, 2020).

Campione d'Italia

Campione d'Italia, on the shores of Switzerland's Lake Lugano, is the second enclaved outlier in Switzerland. It is an integral part of Italy, but located 750 meters inside Swiss territory at its closest point, or 20 km by car (Fig. 9.15). It has a population of around 1960 and is 1.6 km^2 in area (Municipality of Campione d'Italia, 2020; Panzeri, 2019).

Following the collapse of the Roman Empire, the area around Campione was made part of the diocese of Como, but when the

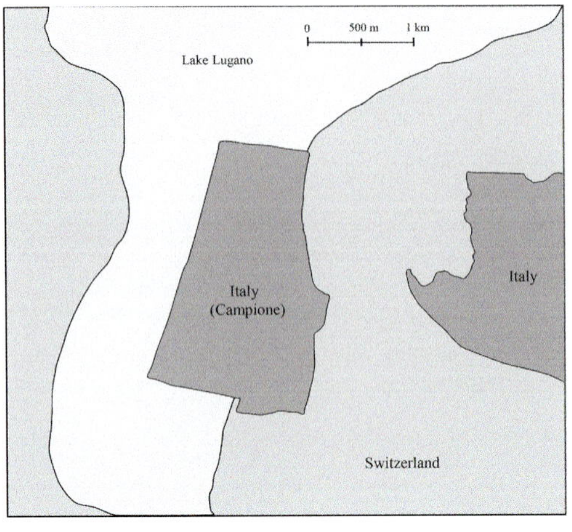

Fig. 9.15. Campione d'Italia's location, enclaved by Switzerland

diocese was divided into parishes, Campione was not included. In the eighth century, the lord of Campione granted its territory as a fiefdom to the Archbishop of Milan, who then passed it on to the Abbey of Saint Ambrose in Milan. For centuries, the Milanese abbot controlled Campione and, when in 1512, the Lugano region was annexed by the Old Swiss Confederacy, Campione became fully enclaved (Catudal, 1979). In 1798, when the Canton Ticino became a full part of the Helvetic Republic and in 1803 the Swiss Confederation, the Campionesi chose to remain part of Lombardy, which was annexed by Italy in 1859.

Campione has been part of the Swiss customs area and economic system since the 1860s. Although it is an Italian town, because of its insularity from Italy, it was not in the European Union and was exempt from the EU's VAT tax regulations. In fact, VAT taxes are not charged at all, and many other taxes are non-existent or substantially lower than in Italy or Switzerland. Being part of the Swiss customs area, there are no passport and customs formalities at its border with Switzerland. Its currency is the Swiss franc, its cars are registered with Swiss number plates and its Italian police drive Swiss-registered police cars (Gardner and Kries, 2005). Switzerland provides telephone service and most of its utilities, and Campioners have access to free health care in Switzerland. In many respects, the enclave's Italian citizens live as though they are Swiss; they are even required to pay duties on items brought from Italy if they exceed their Swiss import allowances (Panzeri, 2019; Vitaliev, 2008).

The tax, employment and social benefits of being a tax-free island in Switzerland ended on January 1, 2020. In 2016, Italy demanded the 'return' of the exclave to the national tax regime, the EU customs union, and back to the effective control of the Italian state. New Year's day 2020 was designated the date of Campione's return. This has been an unpopular move among the exclave's residents, many of whom unsuccessfully petitioned to become part of Switzerland (Economist, 2020). To lessen the blow to Campioners, the local VAT will follow the lower Swiss rate (7.7%) rather than the Italian rate (22%). Also under the agreement, Switzerland will continue to provide services and Italy will pay off its debits to the Swiss. Likewise, a border checkpoint will be established to carry out customs inspections (Euro News, 2020; Municipality of Campione d'Italia, 2020).

Casinò de Campione has long been the tourism lifeblood of the enclave (Gardner and Kries, 2005; Timothy, 1995, 1996). It was established in 1917 to take advantage of its unique location and geopolitics, given that casinos were not permitted in Switzerland or Italy. The casino underwent a major reconstruction in 2007 and employed more of the enclave's population than any other enterprise. Due to mounting debt and economic insolvency, as well as a stronger Swiss franc and Italy's decision to allow slot machines (which Switzerland began to permit in 1993), the casino declared bankruptcy and was closed in July 2018. At the time of writing in 2020, the casino remains closed.

Besides the casino, Campione promotes its museums, historic churches, artworks, entry arch and fountain, beach, and water activities on Lake Lugano (Fig. 9.16). Most visitors are Swiss day-trippers, although there are several hotels for those who choose to stay overnight. With the closure of the casino, arrivals have dropped dramatically. This, together with the lowered competitive advantage by Campione's 2020 transfer to the EU customs union, the community has recently suffered a double-blow to its tourism sector. Despite the closing of the casino and the new customs regime, Campione d'Italia is continuing to target tourism as an

Fig. 9.16. The ornate entrance to Campione d'Italia (photo courtesy of Hugh Wallis)

economic development tool, based largely on its unique border status (Azienda Turistica, 2020).

Llívia

Llívia has the distinct honour of being the oldest of the European exclaves. It was founded by the Romans and was the capital of the Roman region of Cerdanya Valley during ancient times. Much later, the 1659 Treaty of the Pyrenees was signed between France and Spain, ending the Franco-Spanish War (1635–1659). By 1652, Spain had defeated and annexed Catalonia. The Peace of the Pyrenees, as the treaty was called, effectively divided Catalonia between France and Spain and established their border in the Pyrenees Mountains, including the Cerdanya Valley, the location of Llívia. In the agreement, France negotiated to receive 33 villages, including Llívia. However, owing to linguistic differences, Llívia was officially a town or city (*villa*) rather than a village (*pueblo*), so it was not included in the land transfer (Catudal, 1979; Eggers, 2019; Mut Bosque, 2012; Sahlins, 1989), much to the delight of Llívia's population who preferred to remain Spanish and Catalan. In 1866, the home and host states signed a new accord that finally settled the enclave's border (Vitaliev, 2008).

Llívia is situated 1.7 km from the nearest point on the Spanish border. It has a total area of 12.9 km² and a resident population of 1430. It is entirely surrounded by the French *département* of *Pyrénées-Orientales* (Fig. 9.17). More so than any of the other exclaves examined in this section, Llívia is closer culturally, economically and politically to its motherland than the country that surrounds it. The exclave is connected to a secondary road from the borderline at Puigcerdà, the capital of Spanish Cerdanya. This road was built by the French as a 'neutral road' that was subject to French laws but which permitted uninterrupted travel between the Spanish mainland and its exclave (Fig. 9.18). Spanish law enforcement, citizens and merchants were allowed to pass through French territory without inspections, only on this road (Catudal, 1979). Today, however, this

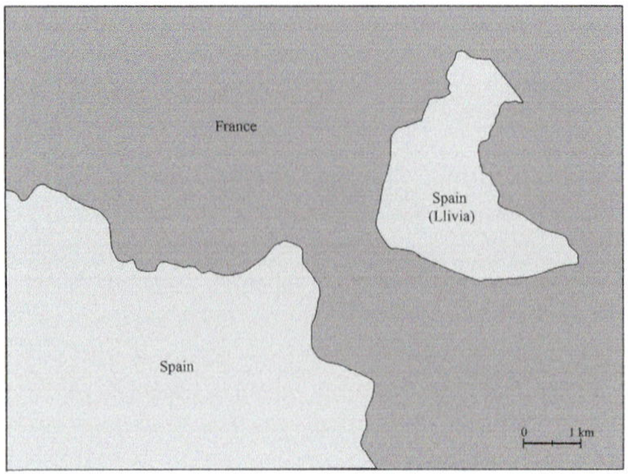

Fig. 9.17. The location of Llívia, Spain, enclaved by France

legal waiver has become redundant as both Spain and France are members of the EU and Schengen. Nearly all public services are provided by Spain, and the enclave has not had dual telephone and postal services, the way so many other exclaves had. As well, its currency has always been the Spanish peseta and the euro after 2002.

The people of Llívia are also more attuned to their motherland and national identity. The Llívienc are extremely patriotic to the motherland—although the motherland is no longer Spain. They are devout Catalonians, and their Catalan identity has strengthened along with that of their compatriots across the border in Catalonia. In the region's October 1, 2017, referendum on Catalonia's independence from Spain, 96% of the Llívienc voted for Catalonian independence. Being an enclave provided certain protections for the community such as in the referendum that was declared illegal by Spain. According to Llívia's mayor, 'The Spanish police were never going to go through France to prevent the people from voting here' (quoted in Fourquet, 2017, n.p.), as they did in other communities in Catalonia. The close social and political connections, ethnic and

Fig. 9.18. Llívia's border with France is marked on the former 'neutral road' by an 1868 boundary stone and white lines on the road

Fig. 9.19. The medieval town of Llívia is now home to many seasonal homes, rental properties, hotels and shops

cultural identity, and economic linkages between Llívia and Catalonia leave no doubt about the loyalty of the Llívienc.

Llívia has seen major tourism growth in recent years with a concurrent boom in construction of primary residences, second homes and commercial lodging establishments (Fig. 9.19). Dozens of hotels and rental homes populate the urban landscape today, and modern buildings have begun to replace the old medieval structures that once characterized its townscape. This is one of the few exclaves with ancient archaeological sites. Roman ruins and the ruins of a thirteen century fortress, as well as the local museum are the main heritage attractions in town. The old pharmacy claims to be the oldest in Europe, having opened in 1594. It is built into the municipal museum (Eggers, 2019). The enclave also hosts the annual international Llívia Music Festival, which is well attended, and in summertime, the area is a popular hiking and cycling destination. In the winter, its hotels and home rentals fill up with skiers, who treat Llívia as a 'bedroom community' for the nearby ski slopes in France.

Despite France and Spain both being in the EU, Llívia's border situation continues to provide economic advantages for shopping, which its French neighbours gladly participate in for significant savings on beer, liquor, tobacco and grocery items, especially the expensive *Jamón Ibérico* (Iberian ham) (Fig. 9.20). In most cases, alcohol and tobacco are taxed lower in Spain than in France (Angus *et al.*, 2019). There are many more alcohol and tobacco shops in Llívia than what can be supported by its own population. According to Vitaliev (2008, p. 240), 'French come here for bargains. Even French wine and French cheese are cheaper here than in France itself'. In terms of tourism promotion, Llívia's marketing efforts are closely aligned with those of Catalonia. Unlike other host state relations

Fig. 9.20. This main thoroughfare in Llívia is home to many shops, permanent dwellings and second homes

with enclaves, French Cerdanya's promotional materials do not generally include information about the enclave.

The Vennbahn German Exclaves

During the 1880s, Prussia built many railway lines, one of which was the Vennbahn. Its original purpose was to carry raw materials (e.g. coal and iron) for industrial production near Prussia's western border. The 1919 Treaty of Versailles altered the German-Belgian border. In 1921, the railroad track, a narrow easement alongside it and its operational structures became part of Belgium. This narrow strip only a few meters wide in some places cut off several small tracts of Germany (totalling 55 km^2) by only the width of the rail line for 28.5 km from Kalterherbert to Raeren (Fig. 9.21). According to the treaty, German marks were the currency of the railway and passengers entering, exiting or transiting on the railroad would remain unchecked by Belgian border officials (EEIG Eifel-Ardennes Marketing, 2020).

The railway was used for various purposes, including tourist sightseeing, until 2002. Most of the tracks were dismantled in 2007 and 2008, and today much of the disused rail bed has been paved over for use as a biking and hiking trail. Although there have been talks in recent years for the rail corridor to revert to German sovereignty, they remain part of Belgium. To the west of the Vennbahn are five small German exclaves, namely Münsterbildchen, Rötgener Wald, Rückschlag, Mützenich and Ruitzhof (Krogh, 2012), which Robinson (1959) termed quasi-exclaves because they essentially do not function like exclaves at all. Because they are only separated by the width of the railroad track and its easement, even during pre-Schengen times, there were no border inspects crossing the line by car, cycle or by foot, and there were no indications of crossing an international border, except

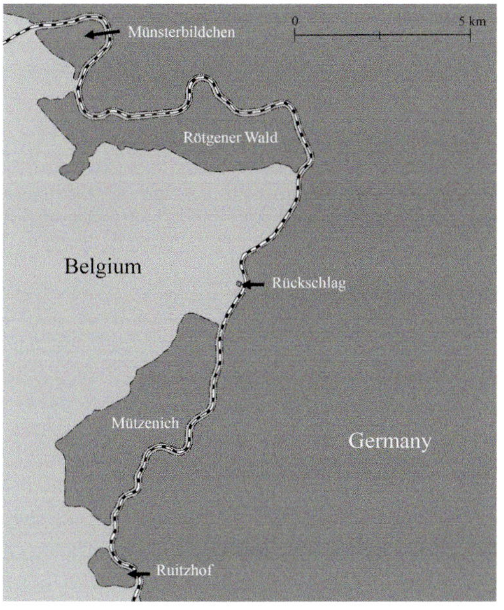

Fig. 9.21. The disused Vennbahn railway line continues to divide Germany from five small exclaves

for the border stones that officially marked the frontier, and Belgian road signs instead of German signs (Fig. 9.22).

Other than for the curiosity factor, the five German exclaves play a very small role in regional tourism. However, the area is scenic and well known for cycling and walking. Four of the five German exclaves have tourist lodging in the form of rural guest houses. By 1990, the line was no longer commercially viable, so communities surrounding it tried to raise money to transform it into a tourist attraction (Vitaliev, 2019), including operating a tourist train. Since its closure in the early 2000s, the Vennbahn has been repurposed into a long-distance cycling route, following the rails-to-trails model established in the United States in the 1970s and 1980s, which aims to transform disused rail corridors into recreational trails (Timothy and Boyd, 2015).

The repurposed Vennbahn itself has become a major recreational resource and tourist route (EEIG Eifel-Ardennes Marketing, 2020). The full trail has been opened since 2013 and

Fig. 9.22. This picture demonstrates the narrowness of the Belgian sovereign territory of Vennbahn; the corresponding border markers can be seen in the foreground and in the paddock behind

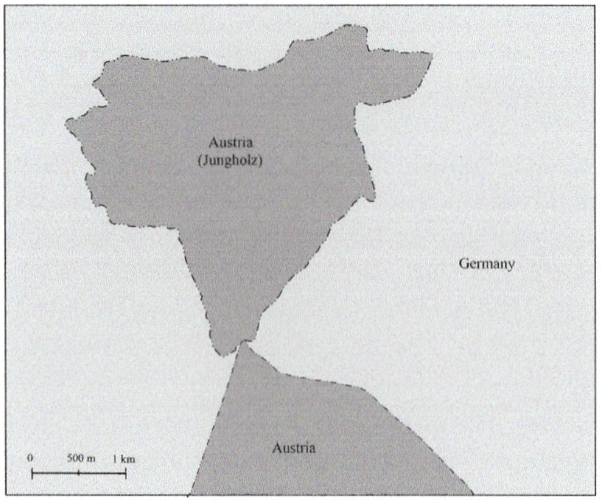

Fig. 9.23. Jungholz is a classic example of a pene-exclave, only accessible through Germany

is an official RAVeL (Autonomous Network of Slow Ways) route, which aims to cater to the recreational and travel needs of families, the elderly, and people with reduced mobility, and other non-racing cyclists (Stoffelen, 2018). Like many tourism development efforts in Europe's cross-border peripheral regions, the project was funded through local collaborative efforts in Belgium and Germany, as well as through generous grants from the EU's Grande Region SaarLorLux INTERREG IV-a program. In total, it is a 125 km leisure cycling route that criss-crosses Belgium, Germany and Luxembourg and is frequently cited as one of the best examples of cross-frontier collaboration in tourism in Europe (Stoffelen, 2018; Stoffelen and Vanneste, 2017).

Jungholz and Kleinwalsertal, Austria

Jungholz is a pene-exclave of Austria. It is attached to Austria at a single point on Mount Sorgschrofen (Fig. 9.23). Given its topography and single-point connection to the motherland, the village can only be visited by going through Germany. Thus, prior to Austria joining the European Single Market and the Schengen Area, Jungholz was included in the German customs area, separated not only physically from Austria but also economically. Remnant of that time, the village still has two postal codes—a German one and an Austrian one, although the postal service is operated by Austria. The enclave is 7.06 km^2 with a permanent population of 319 and an additional population of 94 seasonal inhabitants, who are second home owners (Gemeinde Jungholz, 2020).

Kleinwalsertal (little Walser valley) is another Austrian pene-exclave accessible only through Germany. Like Jungholz, until the mid-1990s it was part of Germany's customs area and was treated as part of Germany in many ways. For instance, German marks were the local currency, and goods brought from Germany could enter tax free, whereas imports from the motherland (Austria) were considered foreign products and were heavily taxed. When Austria joined the Common Market and adopted the euro, Kleinwalsertal's locational tax advantages disappeared. EU efforts to help the area, including Jungholz, under its INTERREG programs have not seen as much success as other cross-border Euroregions have seen (Deppisch, 2008).

The valley is home to three villages, Mittelberg, Hirschegg and Riezlern, with thousands of bed spaces for tourists and approximately 300 lodging establishments. The tourism appeal of these places is less about their geopolitical situations than is the case with other enclaves. Instead they have both become important skiing destinations, with ski slopes, ski schools and ski shops (Vitaliev, 2016). The three Kleinwalsertal villages and Jungholz are scenic representations of the Allgäu Alps and are especially well known for summertime hiking. Kleinwalsertal has extended its product beyond skiing and summer nature walks to incorporate its important agricultural heritage into agritourism and culinary tourism (Kleinwalsertal Tourismus, 2020).

Os De Civís, Catalonia, Spain

This unique and isolated Catalan village is only accessible from the rest of Catalonia through Andorra. A secondary road leads from Sant Julià de Lòria across the Andorran-Spanish border, where Os de Civís lies approximately 2 km inside Spain (Fig. 9.24). The village has a small full-time population. Many of its historic houses are summertime rentals or owned as second homes by Dutch, Spanish and other European investors. The village is home to a Romanesque

Fig. 9.24. The Catalonian/Spanish historic village of Os de Civís can only be accessed through Andorra

church and a small unstaffed museum. It has a couple of hotels, restaurants, and souvenir shops. All of the village's products are imported from Andorra, and the majority of its visitors are day-trippers from other parts of Europe who are spending part of their holidays in Andorra. Its unique medieval villagescape and proximity to Andorra la Vella, make the village an important side-trip destination from the microstate next door.

Samnaun, Switzerland

Samnaun is an example of a relict pene-exclave where, because of its isolation and its only access being from Austria, it was excluded from the Swiss customs area in 1892 and maintained a tax-free status thereafter. In common with similar locations elsewhere, it became a smuggling haven until the 1930s, which became part of its folklore and regional identity. The Samnaun Valley was a full pene-exclave until a road was built in 1905 connecting it to other parts of Switzerland. While in most instances that would have rendered it includable in the Swiss customs area, Switzerland decided to allow Samnaun to retain its tax-free status as a means of economic development (Rougier and Sanguin, 1981).

Today, when crossing from Austria directly into Samnaun, there is no customs checkpoint on the Swiss side, as Samnaun still lies outside the country's VAT and import tax zone. Austria maintains a customs office on its side where standard EU duty allowances apply when returning to Austria. Because of its duty-free status, residents of Samnaun and tourists traveling by road to Martina, the nearest town in Switzerland, even though they remain in Switzerland, are required to pass through Swiss customs upon entering Martina. Because of its duty-free local economy, cross-border shopping from Austria and other parts of Switzerland is a year-round activity. Swiss residents traveling from Samnaun to other areas of the country must abide by Swiss import rules and laws.

Fig. 9.25. The Saimaa Canal on the Russian side of the border

The valley is best known for lower gasoline prices and luxury items (Bechtel, 2004). Prices in Samnaun are more than 15% less than other parts of Switzerland and as much as 30% less than in Austria. More than 50 duty-free shops in a community of less than 800 inhabitants reveals a great deal about the importance of tax-free shopping (Graubünden Ferien, 2020).

Although shopping makes up approximately 65% of the local economy and is popular all year long, Samnaun is a well-known ski resort and winter sport center. The valley shares a cross-border ski area with Ischgl, Austria; together they form one of the largest ski areas in the eastern Alps. The valley is home to many resorts, hotels and guest houses, as well as a multitude of restaurants, cafes and bars. Besides the wintertime buzz, Samnaun is an important summertime destination with more than 250 km of marked hiking trails, mountain biking opportunities, and mountaineering (Graubünden Ferien, 2020).

Livigno, Italy

Livigno is not even a full pene-exclave, as it is connected to the rest of Italy by road. In the summer, the road provides reasonable access, but during the rest of the year, Livigno is essentially a semi-exclave, given the difficulty of navigating the sole mountain road to other parts of Italy. The valley is connected to Switzerland by two other roads, where the Swiss maintain customs checkpoints. Because of this difficult access, winter isolation, and as a way of helping this extreme peripheral area of the country survive economically, Livigno was designated a duty-free and VAT-free zone in 1910, a position that was reaffirmed by the EEC in 1960 and continues today (Christiansen and Smith, 2008; Kovova *et al.*, 2018; Rougier and Sanguin, 1981). The Livignaschi themselves live essentially a tax-free life, except for property and income taxes. They, like all Italians, are permitted to shop tax-free in Livigno, but there is an Italian customs station approximately 15 km down the mountainside. At that point, Italian consumers are subject to the EU's customs regulations, reflecting the same conditions as in Samnaun, as described above.

Today, the area's economy is based overwhelmingly on tourism. Duty-free shopping, winter sports, especially skiing, and summer mountain activities are the foundations of the industry there. The area is home to hundreds of hotels, ski lodges and guest houses. Duty-free shops and petrol stations are a ubiquitous part of the tourism landscape. For a community of only 6400, the number of gas stations and duty-free shops is astounding, which indicates the role of these tax-free activities in the local economy; people come from other areas of Italy and Switzerland to fill their cars, and the duty-free retail outlets employ more than half of Livigno's population (Vitaliev, 2008). Plans for the Milan 2026 Winter Olympics include Livigno for free-style skiing and snowboarding (Azienda di Promozione e Sviluppo Turistico, 2020).

Extraterritorial Places

Extraterritoriality is a subject of international law and a topic studied by political geographers and legal analysts for many years. The concept refers to conditions where entities (certain organizations and foreign representatives) are exempted from the authority and laws of the state or place where they are located. Extraterritorial rights are negotiated bilaterally or multilaterally between individual sovereign states or with multination alliances through diplomatic mediation. This legal condition allows governments to exercise authority beyond their legal boundaries (Rosenthal and Knighton, 2017), so that they can conduct state business abroad without the interference of local laws or foreign meddling. It has most frequently manifested historically in diplomatic immunity for certain accredited representatives of foreign states and historically among prominent global trading companies (Miles, 2019). During the past couple of centuries, it has increasingly been applied to specific places, most notably foreign embassies and headquarters of the United Nations, as well as certain UN agency offices and ships plying international waters (Parrish, 2019).

Saimaa Canal

Although not a normative example of extraterritoriality, the relationship between Finland and Russia as regards the Saimaa Canal emanates many characteristics of extraterritoriality. The cross-border Saimaa Canal is 43 km in length and 34–55 meters wide, with eight locks, three in Finland and five in Russia (Väylävirasto, 2020) (Fig. 9.25). At the time the Saimaa Canal was constructed, 1845–1856, it was entirely within the borders of Finland. The canal was built to provide better transport between the lakes of eastern Finland and the Finnish city of Viipuri (Vyborg) on the Gulf of Finland. As a result of the Second World War, however, Finland lost most of its Karelian region, including Viipuri and nearly half of the canal to the Soviet Union.

The Saimaa Canal is an important means of transporting forest products and other goods between Finland and Russia. It is also a major leisure corridor and heritage attraction, especially for Finns, given its historical connotations for Second World War territorial losses (Timothy and Boyd, 2015). Boats from any country are permitted to use the canal. Russian visas are not needed for transiting through Russian territory on the waterway, but passport inspections are required for recreational boaters at Pälli, the first (or last depending on direction) lock in Russia and at the Nuijamaa Lake checkpoint in Finland. Passengers on organized cruises between Lappeenranta, Finland, and Vyburg, Russia, are inspected in both directions at the ports of Lappeenranta and Vyburg. As the canal was still needed for access to the Gulf of Finland and market access to the Soviet Union, a major trading partner, the portion of the waterway east of the border was leased from the USSR by Finland in 1963 for 50 years. As much of it remained unused and derelict, in the 1960s, it underwent major renovations, reconstruction and restoration with large Finnish investments. The lease was re-negotiated in 2010 with Russia, and several of its legal provisions were clarified. The terms of the treaties clarify that the canal on the Russian side of the frontier is Russian territory, but the canal and a narrow strip of land along its bank are treated de

Fig. 9.26. A Russian border obelisk demarcates the border between the leased canal area and Russian-controlled territory

facto like Finnish territory (Timothy and Boyd, 2015). Finland's Ministry of Transport and Communications maintains the locks, bridges and roads on Russian soil in the leased zone, enforces maritime law, and oversees the traffic flows in the canal, as though it were part of the Finnish transportation system. There are very strict statutes and rules pertaining to private boat owners regarding mooring and stopping in the canal, and wandering beyond the borders of the leased area (Fig. 9.26).

Vatican City's Extraterritorial Properties

According to the 1929 Lateran Treaty between Italy and the Holy See, 23 localities outside the Vatican City state belong to the Holy See, most of which were granted extraterritorial status by the Italian state (Central Intelligence Agency, 2020). These are not counted in the geographical calculations of the Vatican's territory, because they are all an integral part of Italy. Yet, Italy has granted extraterritorial rights to many of these, so that the Holy See can administer them without interference from Italian forces. In essence, they are treated the way foreign embassies in Rome are treated. Several of these properties play an important role in Rome's tourismscape. For example, the Archbasilica of Saint John Lateran with the adjoining Lateran Palace is the location of the signing of the Lateran treaty that created the Vatican City state and is an important tourist attraction in Rome. Likewise, a handful of churches and basilicas fall under this geopolitical protective framework and are also important religious attractions in the city. Examples include the Basilica of Saint Mary Major, the Basilica of Saint Paul Outside the Walls and the Papal Palace and Gardens at Castel Gondolfo, the summer home of the popes and also an important attraction for many people outside of Rome.

Sovereign Military Order of Malta

The Sovereign Military Hospitaller Order of Saint John of Jerusalem, of Rhodes and of Malta, or the Sovereign Military Order of Malta (SMOM) for short, is a unique entity in international law and diplomacy. Some observers argue that SMOM is the smallest sovereign state in the world, even slighter than the Vatican City (e.g. Redazione ANSA, 2016). Yet, the general consensus is that, according to international law, it is not a country or state because it lacks sovereign territory, a permanent population and general recognition as a state by the world community (Cox, 2006; Klieger, 2013; Stojanovska-Stefanova and Atanasoski, 2016). It is, in the words of Farran (1954, p. 222), a 'persona mixta between a sovereign state and a religious order'. Stojanovska-Stefanova and Atanasoski (2016) describe it as an organization and a de jure government, rather than a de jure state.

SMOM has an interesting story. By the eleventh century, Jerusalem had become a significant pilgrimage destination for Christians from Europe and the Middle East. A convent and hospital were built in the early eleventh century in Jerusalem, where the monastic Order of St John of Jerusalem cared for the wellbeing of Holy Land pilgrims. By Papal decree, the pilgrimage center was beholden to the Holy See rather than the local church; under the laws of that time, this charter essentially granted the organization sovereignty and autonomy to function outside the control of the civil authorities of its location. Under the Crusader Kingdom of Jerusalem, the order's Knights of Hospitallar led a care-taking role, while the Knights of the Templar were responsible for military defense. With the defeat of the Kingdom of Jerusalem in 1291, the order move to Cyprus and several years later, to Rhodes where it remained for 213 years as the ruling power with a formidable naval force charged with defending the Christian world and protecting and serving pilgrims. In 1523, the Ottomans expelled the order from Rhodes, displacing the sovereign entity, which remained stateless (without its own territory) until 1530, when the King of Sicily gifted the Maltese islands to the Order of St John. The order was the government of Malta until 1798, when Napoleon's forces seized the Maltese archipelago and banished the Knights of Malta. This was the last time the Sovereign Military Order of Malta possessed its own territory.

Fig. 9.27. Palazzo Magistrale, the 'capital' of the Sovereign Military Order of Malta

After temporarily settling in various localities in Italy, the organization dropped much of its military prominence, re-emphasized its hospitallar origins to provide humanitarian aid, medical assistance and social support, and moved to Rome in 1834. Today it occupies two buildings that have been granted extraterritorial status and have been in possession of the order since the seventeenth and fourteenth centuries. These are the *Palazzo Magistrale* (Magistral Palace) near the famed Spanish Steps and the *Villa del Priorato di Malta* (Magistral Villa) on the Aventine Hill, 2.5 km apart (Fig. 9.27).

SMOM functions today as an international Catholic religious order, which provides aid and medical care for the world's needy in 120 countries, and has recently been involved in providing for the needs of refugees and migrants arriving in Europe, and dealing with the global COVID-19 pandemic. Its remaining Military Corp facilitates paramedic functions and assists the Italian army in its care for the injured.

Fig. 9.28. The Spanish embassy to the Holy See and SMOM

As it has throughout its history, SMOM mints its own coins in its official currency, the scudo, although it uses the euro for its everyday functions. It has its own postal system and produces its own car license plates, which can be seen throughout Rome. As a sovereign and independent religious order, it is not bound by Italian law within the walls of its two buildings or with regard to its humanitarian functions. According to its constitution, it conducts its own internal and foreign affairs wihout the intervention of any state (Farran, 1954; Stojanovska-Stefanova and Atanasoski, 2016; Timothy, 2003). Also as a sovereign body, it issues passports, which are accepted by countries throughout the world, and has established diplomatic relationships with 110 countries. It maintains embassies in most of these countries, and many countries have established embassies to SMOM in Rome, whereas some countries' embassies to Italy or the Holy See are accredited simultaneously to SMOM (Sovereign Order of Malta, 2020) (Fig. 9.28). It also has permanent observer status in most United Nations agencies, including the UN itself, FAO, UNESCO, and WHO.

Beveridge and O'Gorman (2013) shed light on how the forerunners of the Order of Malta played a key role in ancient tourism and hospitality. The knights (Hospitallers) provided accommodation and shelter, food and medical care for pilgrims during medieval times in Jerusalem and security (Templars) for pilgrims and other Christians in the Holy Land, Rhodes and Malta, as well as elsewhere in Europe. 'This saw the beginnings of organised mass movement of travellers to the "Holy Lands" under the collaboration of the Templars and the Hospitallers, offering security, shelter and safety while travelling to an occupied land during turbulent and dangerous times' (Beveridge and O'Gorman, 2013, p. 44).

Although the organization's mandate today is related to medical care and rescue, it is assuredly involved in tourism in a variety of ways (Timothy, 2003). The organization's main administrative offices and post office are located in *Palazzo Magistrale*. At the post office, Romans can post their mail and tourists can buy SMOM stamps and send postcards. Although curiosity-seekers are not permitted entry into the main gate and administrative area of the building, SMOM's website encourages tourists to visit the 'newly-opened Visitors Centre [which] provides curious passers-by with an opportunity to learn more about the Order of Malta' (Sovereign Order of Malta, 2020, n.p.). The simple visitors center provides interpretive information about the history of the order and its aid work in 120 countries (Fig. 9.29).

Villa Magistrale houses SMOM's embassy to Italy and hosts many international events and delegations. It is an hisotoric ediface of significant cultural importance, although tourists are only allowed to visit by reservation one morning each week. The main attraction of the villa is the *Buco della Serratura*, a keyhole through which visitors can see one of the most iconic views in all of Rome: the dome of St Peter's Basilica framed with the trimmed hedges of SMOM's villa property (Sovereign Order of Malta, 2020).

In addition to its two extraterritorial headquarters, SMOM is an active promoter, organizer, and helper of international pilgrimages. The order has for many years organized Catholic pilgrimages to sacred sites in the Holy Land, throughout Europe and even in North and South America. Getting back to its original roots 900 years ago in Jerusalem, the organization also operates ambulance services and first aid and rescue stations in the major pilgrimage destinations

Fig. 9.29. The entrance to SMOM's visitor information center, post office and hospital in Palazzo Magistrale

of Medjugorje, Telgte, Knock, Croagh Patrick, Guadalupe, Fátima and Santiago de Compostela. The Order of Malta has operated the relief station at the final leg of the Camino de Santiago for 20 years (Sovereign Order of Malta, 2020).

Conclusion

Most people have never heard of the enclaves and extraterritorial areas of Europe. These do not stand out as major international tourism destinations, but they are significant attractions in their local regions. Exclaves, like microstates, are heavily dependent upon tourism. Their unique borderland locations and the legal loopholes that have enabled locals to survive for centuries have now become mechanisms to further the cause of tourism. The most common enclavic manifestation of this on a local level is in the area of shopping. For a more global audience, it is their 'curiosity factor' that tends to attract people who hear about them in travel reporting or in news stories. True exclaves, pene-exclaves, and counter-exclaves all have the same appeal: their separateness and unique local identities that are molded by years of border restrictions and opportunities on the margins of two states.

Extraterritorial areas are similar to exclaves in that the laws that surround them have no direct impact upon the functions of ordinary life. While some of these are of themselves unremarkable in offering typical tourism advantages (e.g. SMOM), others have bountiful historical, natural and heritage value (e.g. Saimaa Canal and the Vatican's outliers) that attracts recreationists, sightseers or worshippers. What they all have in common, exclaves and extraterritorial entities alike, however, is their curiosity factor.

10

Unrecognized Polities: The De Facto States and Micronations of Europe

Introduction

The geographical anomalies examined so far in the book all have one main characteristic in common: they are recognized and accepted by the global community as legitimate sovereign states or integral parts of sovereign states. Yet, there is still a classification of places in Europe that defies the patterns that have been discussed so far. These are termed here 'unrecognized polities', or locations that claim to be sovereign states but do not exist legitimately outside of their creators' own imaginations or local realms of power. Although they do occupy physical space and have socio-economic functions and unique identities, and are therefore real places (Relph, 1976), they are not universally accepted as legally defined sovereign states. In spite of their lack of universal recognition, the two types of entities examined in this chapter (de facto states and micronations) are keenly interested in targeting tourism as both an economic development driver and a propaganda stratagem.

De Facto States

A de facto state is an area of a legitimate (universally recognized) state that has seceded and functions as though it were an independent country. What distinguishes these from newly formed sovereign states is the way in which they came about (nearly always through conflict) and whether or not they are acknowledged by the world community as independent states. Many terms have been used to describe de facto states, such as quasi-states, proto-states, pseudo-states, unrecognized states, breakaway states, states within states, shadow states, and states without sovereignty (Caspersen and Stansfield, 2011; Souleimanov, 2020). Although there are subtle differences between some of these concepts, they effectively refer to the same phenomenon and will be used interchangeably in this chapter.

There are currently seven de facto states in Europe, all in Eastern Europe and the eastern Mediterranean (Table 10.1) (Fig. 10.1). A couple of others have existed in recent decades (e.g. Găgăuzia and the Chechen Republic of Ichkeria) but have dismantled through diplomatic efforts and the granting of greater autonomy within the legal home state. Most of Europe's shadow states came into existence with the dissolution of the Soviet Union in the early 1990s (Hoch, 2020a), although in the case of Northern Cyprus, the situation occurred much earlier. Most research on de facto states has focused on their origins and causes, external relations, internal dynamics

Table 10.1. Information about de facto states in Europe

Entity	Capital	Area	Population (est)	Legal sovereignty	Foreign patronage	Declared independence	Language(s)	Currency
Abkhazia	Sukhumi	8,653 km^2	246,313	Georgia	Russia	July 23, 1992	Abkhaz & Russian	Abkhazian apsar & Russian ruble
Artsakh	Stepanakert	11,433 km^2	147,906	Azerbaijan	Armenia	September 2, 1991	Armenian	Artskah dram & Armenian dram
Donetsk People's Republic	Donetsk	7,853 km^2	2,302,444	Ukraine	Russia	May 12, 2014	Russian & Ukrainian	Russian ruble
Luhansk People's Republic	Luhansk	8,377 km^2	1,464,039	Ukraine	Russia	May 12, 2014	Russian & Ukrainian	Russian ruble
Northern Cyprus	Nicosia	3,355 km^2	325,890	Cyprus	Turkey	November 15, 1983	Turkish	Turkish lira
South Ossetia	Tskhinvali	3,900 km^2	57,652	Georgia	Russia	December 21, 1991	Ossetian & Russian	Russian ruble
Transnistria	Tiraspol	4,163 km^2	503,640	Moldova	Russia	August 25, 1991	Russian & Romanian	Transnistrian ruble

Sources: Population Data (2020); Central Intelligence Agency (2020)

Fig. 10.1. The locations of Europe's de facto states (Source: basemap from d-maps.com)

(e.g. state-building), and democratic processes or lack thereof (Hoch, 2020a).

All European unrecognized states have at least four political characteristics in common. First, they all formed out of conflict situations (Hoch, 2020b). Second, they are ethnically defined political entities (Kopeček, 2020a). Third, they do their best to imitate the democracies and socio-political systems of their patron state (Caspersen, 2011), and fourth, they all have a strong dependency relationship with their external supporter state (Metreveli and Timothy, 2010; Veenendaal, 2016a). This last point is known as a patron-client relationship, which entails the patron state harboring the separatist entity and supporting it financially, morally and administratively. Such efforts are often done in the name of *irredentism*—actions to protect national minorities living abroad (e.g. Russians living in Ukraine) and to promote the return of national territory to the current nation-state where it used to belong (Glassner and Fahrer, 2004). In return for its patronage, the client state (in this case, the quasi-state) is devoted to the patron, providing benefits to its larger supporter (i.e. master), such as resources or access to a certain region or strategic location. In Kosienkowski's (2020, p. 197) analysis,

> Normally patrons achieve a client's compliance with their needs through inducement, that is, providing and increasing support to clients. The more a given client values gained benefits, the more it is responsive to a patron's demands. De facto states can be classified as clients that highly regard a patron's assistance, given that it ensures their survival. Moreover, this support is hard to replace. The point is that, due to the illegitimate status of de facto states, the number of their potential patrons is extremely limited, if not non-existent....As such, de facto states can be expected to be highly responsive to the demands of their patrons.

With the exceptions of Artsakh and Northern Cyprus, whose patron states are Armenia and Turkey respectively, Russia is the patron country of all of Europe's other unrecognized states, which reflects the fact that it was the collapse of the USSR and the 2014–present Russia-Ukraine crisis that created the majority of them (Daase, 2016; Hoch, 2020b; Hoch and Souleimanov, 2020; Kopeček, 2020a; Šmídová and Šmíd, 2020). Russia's patronage of Abkhazia, South Ossetia and Transnistria reflects its lingering desire to strengthen its influence over the Caucasus, Moldova and Ukraine (Hoch and Souleimanov, 2020). Russia, Armenia and Turkey all have been successful in their irredentist and patronage campaigns to support these pseudo-states. Close relations with the external patrons, the patron state's espousal of legitimacy, and the harboring state's help with incubating the individual national identity of the shadow republics means there is little interest in returning to the parent state (O'Loughlin *et al.*, 2014), although this varies among ethnic groups within certain de facto states, such as among Russians, Ukrainians and Moldovans in Transnistria.

Although differing in a variety of ways among the unrecognized states, tourism exists in all of them and has been considered an important development tool they wish to pursue. The most obvious tourism patterns they all share are that their main market is the patron state and that their history of conflict and self-proclaimed sovereignty is their primary asset. In terms of demand, the main market for Abkhazia, South Ossetia, Transnistria, Luhansk and Donetsk since their establishment has been Russians (German, 2016; Gerrits and Bader, 2016; Kolstø, 2020; Metreveli and Timothy, 2010; Sideri, 2012). Because of its pleasant climate, nice beaches, and natural landscapes, Abkhazia was already a popular tourist destination for Russians during Soviet times—a position that continues to the present day (Gumba and Ketsba, 2004; O'Loughlin *et al.*, 2011; Popescu, 2006a), although the past few years have seen a decline in Russian arrivals owing to increased violence against Russian tourists (Kolstø, 2020). To sustain this tourism connectedness between the client and patron states, Russia has invested a great deal in Abkhazia's tourism industry, from transportation infrastructure to accommodations and beachfront development (Cooley and Mitchell, 2010; Gerrits and Bader, 2016; O'Loughlin *et al.*, 2011). Russia has made considerable investments in the tourism sector in all of its client states (Popescu, 2006b).

As noted above, from a supply perspective, the history of conflict that created the de facto states are their main tourism asset (Lyubitseva and Zavarika, 2019). Not only has this put them in the media spotlight, it also creates a sense of mystique and adventure. The people who do visit seek to collect places, to say they have been somewhere unique (Lambert, 2018). The title of one National Geographic travel report speaks volumes about this element of appeal of the unrecognized states: 'Visit a country that doesn't exist: Here's why bold travelers should visit Transnistria, a mysterious land in limbo' (Vaello, 2018, n.p.).

For the most part, however, none of them are overly popular destinations beyond the patron-state market, largely because of the conflicts that created them and which continue to maintain them in conditions of unofficialdom. Conflict, even dormant struggles, are a deterrent to tourism (Lyubitseva and Zavarika, 2019). The individuals who do visit, are a small niche comprising of people who are interested in political history and in seeing the place for themselves. The sections that follow describe three de facto states in Europe and highlight some of the tourism trends and issues they face.

Artsakh

The Republic of Artsakh, sometimes known as the Republic of Nagorno-Karabakh, is an area of Azerbaijan that declared independence from the Soviet Union in 1991. During the Soviet era, Nagorno-Karabakh was an Armenian ethnic island within Azerbaijan near the border of Armenia. When the USSR dissolved, the Nagorno-Karabakh Autonomous Oblast declared independence with the aim of unifying with the new sovereign state of Armenia. Azerbaijan immediately rejected the pronouncement, which led to the Nagorno-Karabakh War until 1994 (Cornell, 1999). Out of this conflict, the Republic of Artsakh arose and included Nagorno-Karabakh and some additional territory that connects the quasi-state with the Armenian homeland physically and serves to provide a buffer with Azerbaijan.

The global community recognizes the area of Artsakh as a legitimate part of Azerbaijan, but the breakaway republic functions as a separate country with its own government and administrative services, albeit with a heavy dependence on financial and administrative assistance from Armenia. Although irredentism ostensibly lies at the heart of the patron-client relationship between Armenia and Artsakh, the former's support for the Karabakh secessionist effort provided strength for its own struggle for emancipation from the USSR and its own legitimacy on the global stage. Once these were established, a more integrative approach was formalized (Kopeček, 2020b).

This cultural homogeneity of the proto-state and its Armenian motherland, the external threats posed by Azerbaijan, the strength of solidarity among the Armenian diaspora, and the continuous state of non-recognition have strengthened the resolve of Artsakh and helped it become one of the most successful shadow states in Europe (Kolstø and Blakkisrud, 2012). Whereas in most cases, a lack of official recognition and conflict with the legal home state causes foreign investors to shy away from financing projects in de facto states, China has invested relatively heavily in Artsakh's tourism in recent years (Lambert, 2018), supplementing the development funds that derive directly from Armenia.

Artsakh's unrecognized government has labored to develop the country's tourism infrastructure and promote the pseudo-state as a special destination. Built heritage is Artsakh's primary tourism asset; like its neighbors Armenia and Georgia, it has some of the oldest Christian churches and monasteries in the world. Thus, there is significant potential to develop pilgrimage tourism and other forms of heritage tourism (Lambert, 2018). The breakaway state is also on the cusp of developing a wine tourism sector. As noted earlier, the majority of tourists in Europe's de facto states come from the patron states. This is true in Artsakh's case, where nearly all tourists are Armenians, but there is also a strong showing of people from the Armenian diaspora throughout the world (Crowley, 2013). Diasporic Armenians often come to show solidarity with the breakaway country, to support its independence, or in Artsakh's case, its desire to reunite with the Armenia motherland. All travel to Artsakh takes place via Armenia and Armenian travel agents. The unrecognized state's development efforts have largely been done in collaboration with the tourism efforts of Armenia. The breakaway country's infrastructure is poor, although there are serious discussions about developing it, including re-opening the derelict airport in the capital Stepanakert.

The early 1990s war, the current stalemate with Azerbaijan, and Armenia's support for Artsakh's incipient independence brought about a trade and travel embargo against the proto-state and its motherland by Azerbaijan and Turkey (Cappucci and Zarrilli, 2008). Technically, Armenia and Azerbaijan are still at war, and the April 2016 Four-Day War between them escalated tensions once again. The breakaway state had seen a growth in tourism until April 2016, but that short-lived conflict caused visitation to collapse, although a resurgence of tourism was noticeable after 2017. As of mid-2020, tourism continues to grow from Armenia and from abroad, even though Armenia's borders with Turkey and Azerbaijan remain closed. Citizens of Armenia are not permitted to enter Azerbaijan, and anyone possessing evidence (i.e. passport stamp) of having visited Artsakh will be refused entry as well.

Relations between Armenia and Azerbaijan have the potential to improve in the next few years. Since 2018, there have been hints of renewed talks between the two states, but Armenia's insistence that the Nagorno Karabakh region be represented in the talks under the name of Artsakh appears to have riled the Azerbaijanis and stalled the negotiations and in July 2020, armed clashes resumed. With the exception of the other post-Soviet de facto states of Transnistria, South Ossetia and Abkhazia, no other sovereign state in the world recognizes Artsakh's independence.

Transnistria

Transnistria, or more officially the Pridnestrovian Moldavian Republic (also known as Pridnestrovia or Transdniestria), is the portion of Moldova that lies between the Dniester River and the Ukrainian border. In 1989, the Soviet Republic of Moldavia replaced Russian with Moldovan (Romanian) as the only official language, which upset the Russian and Ukrainian population of eastern Moldova and caused considerable contention between the republic's ethnic groups. In 1990, in response to perceived maltreatment and marginalization by the Moldovan majority, Transnistria announced its unilateral separation from Moldova as the new Pridnestrovian Moldavian Soviet Socialist Republic, and while this was not accepted by the Soviet Union or Moldavian authorities, Transnistria's governance system formed separately, and the territory began to function independently from the Moldavian Soviet Socialist Republic (SSR).

The Moldavian SSR declared independence from the Soviet Union (as Moldova) on August 27, 1991. With the final dissolution of the USSR on December 26, 1991, the Transnistrian issue lost the attention of Moscow, although various supporters from Russia and Ukraine came to Transnistria to fight against Moldovan efforts to rein in the rebellious region. Armed conflict between Moldovan forces, Transnistrian forces, and foreign volunteers for the Transnistrian cause lasted until July 21, 1992, when a ceasefire agreement was signed. The ceasefire continues to hold, and the Transnistrian issue remains unresolved (Kosienkowski, 2020). Although the global community recognizes it as a *de jure* area of Moldova, the Pridnestrovian Moldavian Republic (i.e. Transnistria) operates as an independent country, providing the normal services (e.g. education, transportation, security, health care, border controls) that sovereign states generally provide.

Relations between Russia and Transnistria are an archetypal patron-client state relationship. Russia benefits by having a physical presence in a strategic location vis-à-vis Moldova and Ukraine. Although Transnistria's desire is an eventual unification with Russia, Russia desires the breakaway territory's reunification with Moldova as an avenue to increase its power in competing with the West for influence in the former Soviet republic. Although it borders Ukraine, Transnistria does not have good relations with its eastern neighbor given Ukraine's ongoing conflicts with Russia since 2014. As the client state, in return for Russia's patronage, Transnistria receives protection for its Russian population, internal security, economic support, backing for domestic affairs, and diplomatic representation abroad (Kosienkowski, 2020). In reality, without Russian patronage, the unrecognized Pridnestrovian Moldavian Republic would cease to exist.

Although in most ways, Transnistria functions as a separate state or puppet state of Russia, Moldova's sovereignty overrules certain aspects of international relations. For instance, Transnistrian companies trading with Ukraine must be registered as Moldovan companies. Although the country issues Transnistrian passports, these can only be used for visiting the other ex-Soviet breakaway states (i.e. Abkhazia, South Ossetia, and Artsakh). Transnistrians must travel abroad using Moldovan, Russian or Ukrainian passports; dual and triple citizenship is common. Elections in Transnistria are not considered valid by Moldova, the EU and other international observer bodies.

In common with its Soviet forebear, Transnistria has pursued a heavy industry-based economic development model (Lambert, 2018). Although still seriously underdeveloped in most respects, tourism is becoming an increasingly important part of the Transnistrian economy (O'Loughlin *et al.*, 2014). Although most tourists in Transnistria come from Russia, other nationalities are visiting in increasingly higher numbers, especially the Chinese. All international arrivals come by road or rail, as the nearest airports are in Chişinău, Moldova, or Odessa, Ukraine. There are Transnistrian checkpoints on its borders with Moldova and Ukraine.

Transnistria is known among adventurers as an off-the-beaten path destination where people can enjoy a taste of the old Soviet Union, because much of the old country's iconic communist symbols (e.g. hammer and sickle emblems), monuments (e.g. statues of Lenin), architecture, and everyday life (e.g. old Soviet trolleys) have been relatively well preserved in a nostalgia-inducing Soviet bubble (Lambert, 2018). Even though Transnistria is not a communist state, it values its Soviet past and nostalgically preserves the remaining Soviet landscapes. Lambert (2018, p. 119) observes that Berlin-based 'Soviet Tours', a new company, even offers a whole trip to all de facto states for curious adventurers. As tourism and 'hipster' culture is developing, more people are going to Moldova and feel like they must go to Transnistria to enjoy a few days in the 'land of the last Soviets'. This is reminiscent of the contemporary communist tours available in many Eastern European cities, which guide visitors through the region's socialist past, visiting factories, secret police headquarters, Soviet-style apartment blocks, and eateries (canteens), all while being transported in an East German Trabant or Soviet Lada (Timothy, 2011). These old-time communist tours in Transnistria and other post-Soviet de facto states cater to political history enthusiasts and Europeans who lived through the communist era and yearn to experience some level of wistfulness of that historic period.

Turkish Republic of Northern Cyprus

Of all the de facto states in Europe, tourism is best developed and most researched in the Turkish Republic of Northern Cyprus (e.g. Alipour *et al.* (2011); Altinay and Bowen, 2006; Hayat, 2019; Mehmet, 2010; Seddighi *et al.*, 2002; Sharpley, 2001, 2003; Yasarata *et al.*, 2010). Cyprus is a partitioned state, divided by a de facto international boundary, which is not recognized by any other state in the world besides Turkey (Butler and Mao, 1996; Webster and Timothy, 2006) (Fig. 10.2).

Following years of post-independence contention between the island's Greek and Turkish populations, and in response to a coup d'état by insurrectionists who favored unification with Greece, Turkish military forces arrived on the island in two waves in July and August 1974, occupying the northern third of the island. In August, a ceasefire line was drawn from west to east, partitioning the island into two areas: the Greek-populated south and the Turkish-populated north. To achieve demographic segregation and to avoid further ethnic clashes and maltreatment, most Greek Cypriots in the north moved to the southern portion of the island, and the Turkish Cypriots migrated northward. Approximately 185,000 Greek Cypriots

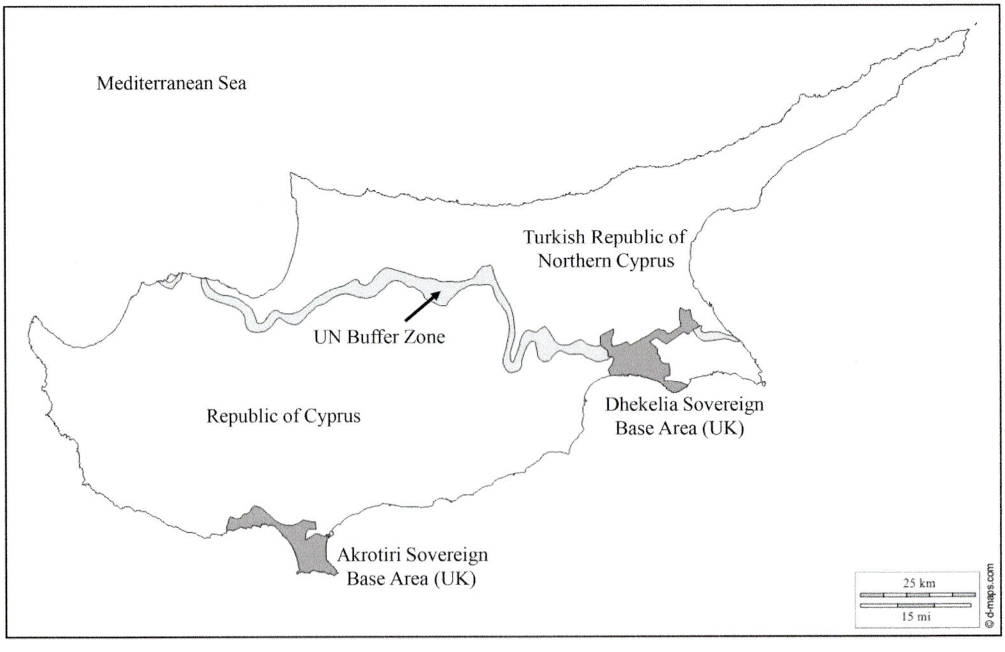

Fig. 10.2. Divided Cyprus: the Turkish-controlled north, Greek-controlled south, the UN Buffer Zone and the two Sovereign Base Areas (Source: basemap from d-maps.com)

moved to the south, while some 45,000 Turkish Cypriots moved across the ceasefire line into the north (Kliot and Mansfeld, 1994). During the 1974 conflict, thousands of Greek and Turkish Cypriots disappeared or were killed on both sides. Since that time, relations between Greek and Turkish Cypriots have been severely strained, as well as the relationship between Turkey and Greece (Leigh and Vukovic, 2011).

In 1974, United Nations peacekeeping forces arrived to enforce the ceasefire agreement by establishing a 180km demilitarized buffer zone between the Greek and Turkish sectors of the island (Souter, 1984). Inter-sectoral travel ceased immediately, and the progress in tourism that had been made on the island up until that time halted entirely. The Green Line, as the new border was known, sliced through the island, severing not only pan-insular tourist flows but essentially ending tourism entirely (Timothy, 2001c).

In the 1960s and early 1970s, tourism had become a salient part of the Cypriot economy, and the island had developed into a desirable winter destination for British and northern Europeans. Most development had occurred in the northern part of the island, particularly in the coastal cities of Famagusta and Kyrenia, where two-thirds of the island's accommodations and tourist arrivals were concentrated (Timothy *et al.*, 2004) (Fig. 10.3). Following Turkey's invasion (or intercession from the Turkish perspective), an economic embargo was enacted against the Turkish-occupied part of the island by the international community; all foreign visits were prohibited except by Turkish citizens, and Turkey was the only country to carry out trade with the north.

On November 15, 1983, the occupied zone declared independence from Cyprus as the Turkish Republic of Northern Cyprus (TRNC) and has operated as a de facto state since that time. The country's lack of recognition and the trade and travel embargo severely hindered the growth of the TRNC economy (Kliot and Mansfeld, 1997; Mehmet, 2010; Mehmet and

Fig. 10.3. Kyrenia Harbor today. Kyrenia was one of Cyprus' most popular destinations prior to 1974

Tahiroglu, 2002). Because of these conditions in the north, tourism stalled, although a small tourism industry developed based on Turkish arrivals (Mansfeld and Kliot, 1996). Because the island had significant tourism potential and because the international community empathized with the Republic of Cyprus, billions of dollars were invested in developing tourism in the southern part of the island, with the result that tourism there developed rapidly and thrived throughout the 1980s and 1990s, especially as a winter destination.

TRNC's tourism, and economic development in general, lagged far behind that of the Republic of Cyprus for a number of reasons. First, Turkey and its client state, TRNC, were viewed by the world as the aggressors in the conflict, which resulted in a general sense of insecurity and sympathy for the Greek Cypriots. Second, even if foreign tourists wanted to visit the north, they were generally prohibited from doing so because of the sanctions on the north that prohibited foreign direct investments and direct tourist arrivals. Third, because of the first two issues, the Turkish officials focused their development efforts on agriculture and manufacturing rather than tourism (Timothy *et al.*, 2004).

After the 1974 invasion, the island's two governments strictly proscribed cross-border travel. Whereas Cypriots were prohibited from crossing in either direction, foreign tourists were allowed to visit the north during the 1980s and 1990s in one of two ways: directly through Turkey or on day-trips from the south. Anyone entering Cyprus in the north (i.e. from Turkey) was considered an illegal entry, and the person was barred from entering the Republic of Cyprus. Foreign tourists were permitted to cross the Green Line on same-day return trips, as long as they had legally entered Cyprus in the south, crossed only during the daytime, did not stay overnight in the north, and they did not bring back any items purchased in the north (Timothy, 2001; Webster and Timothy, 2006).

On April 23, 2003, conditions for Cypriots changed dramatically when TRNC suddenly and unexpectedly lifted restrictions on Greek Cypriots visiting the north and Turkish Cypriots visiting the south. In preparation for joining the EU the following year, the Republic began allowing the same transfrontier mobilities in 2003

Fig. 10.4. Like St Nicholas Church, many churches in the north were converted into mosques after 1974

that the TRNC had instigated. Thus, all Cypriots were unexpectedly authorized to visit both sides of the island (Díaz-Sauceda *et al.*, 2015; Webster and Timothy, 2006). This immediately led to huge queues at border crossings on both sides of the demilitarized zone and the subsequent opening of a handful of new border crossings (Timothy *et al.*, 2004).

Permission to cross resulted in hundreds of thousands of border crossings. The most eager crossers were those who had lived on the other side but were forced to relocate as refugees in 1974–1975. Greek Cypriots headed north to see what life was like in 'enemy-controlled' territory and to visit their family farms, homesteads and churches (Fig. 10.4). Turkish Cypriots headed south to visit their former homes, find old friends, and see the community mosques left behind (Webster and Timothy, 2006). As time passed, border crossings normalized, and Greek Cypriots frequented the north to play in the casinos that had opened there in previous years (Fig. 10.5). Many Turkish Cypriots crossed daily for shopping and to work in unskilled, low-cost labor opportunities in the south. In the early days of the

Fig. 10.5. Casinos became popular in the Turkish north once the border was open to all crossings in 2003

border opening, many Greek Cypriots refused to enter the de facto state in the north in principled defiance of the Turkish occupation, including a refusal to show their passports to TRNC officials in what they considered to be their own country (Webster and Timothy, 2006).

Today, it is still illegal for foreign tourists, including EU citizens, to enter the island in the occupied north, and the Republic government may take punitive actions against illegal entries. Day-trips from the south to TRNC and vice versa have become routine for foreign tourists and Cypriots. Although passports are required, they are no longer stamped, and the Republic of Cyprus inspectors south of the buffer zone typically wave EU citizens through. In addition to the new border checkpoints that opened in 2008 and 2010, two more were opened in November 2018, for a total of nine official border crossings. Two of these are on the border of TRNC and the British Sovereign Base Area (SBA) of Dhekelia, with southbound crossings regulated by the SBA police (Fig. 10.6); the remainder of the crossings are located directly on the edges of the buffer zone. Pedestrians may cross at any of these, and eight are open for motor vehicle crossings. Some car rental agencies in the south will permit drivers to take hired cars across the border, but Republic insurance is not valid in TRNC. All of the border crossings are open 24 hours, and the restrictions on spending the night in the north no longer exist. Likewise, border crossers are now permitted to buy items in the north, although the Republic does impose strict limits on tobacco, alcohol and certain other items.

Fig. 10.6. A border crossing point between the Dhekelia SBA and the TRNC

Although the European Union counts the whole island as an EU member and all Cypriots as EU citizens, the north is considered to be occupied EU territory that currently lies outside the control of the Republic of Cyprus. Thus, for all intents and purposes, EU law is suspended in the Turkish-controlled north, such as the free movement of people, capital, goods and services. The EU's stance is 'state avoidance', to engage informally with TRNC without validating state recognition (Casaglia, 2019; Kyris, 2020).

Cyprus is a relatively small island state (some observers consider it a microstate—e.g. Panke, 2011) that has a well-established tourism industry. It is particularly known for its beaches and resort communities, historic churches and monasteries, heritage towns, archaeological sites, and natural areas. Three of its most important heritage assets are on UNESCO's World Heritage List. Despite its heritage and natural richness, one of the most abiding attractions on the island is the UN Buffer Zone that separates the Turkish Republic of Northern Cyprus from the Republic of Cyprus (Figs. 10.7 and 10.8). In Nicosia, it is an especially important attraction, and crossing over it has become a matter of ritual for the majority of tourists who visit the capital city (Díaz-Sauceda *et al.*, 2015; Gelbman, 2010; Webster and Timothy, 2006).

Ledra Street was one of the most important retail zones in Nicosia prior to the 1974 event. When the ceasefire line was established through the capital city that year, Ledra Street was barricaded on both sides of the buffer zone, creating a 75 meter-wide 'no man's land' between the northern and southern parts of the city at that point. Shortly thereafter, observation platforms were built on both sides to allow locals and tourists

Fig. 10.7. Inside the UN Buffer Zone in Nicosia

Fig. 10.8. The UN Buffer Zone in Nicosia

Fig. 10.9. Before 2007, this observation platform was an important attraction in Nicosia where tourists could view the 'other side' of the island

to look over the buffer zone into the 'foreboding other'. The platform on the Greek side became one of the most visited tourist attractions in the city (Fig. 10.9), but with the 2003 opening of the border, the appeal of the observation points diminished. The Ledra Street wall and platform on the south side were demolished early in 2007. On both sides, Ledra Street was restored, the disarray and dereliction inside the UN buffer zone was cleaned up, and the site was opened again for pedestrian crossings on April 3, 2008.

Micronations

There is another category of non-recognized entity, which has received hardly any academic attention but considerable media attention. It differs significantly from the de facto states described above in legal, political and geographic terms. Micronations—not to be confused with microstates—are self-declared 'countries' that are neither recognized by other states and have no patron state, nor in most cases have any claim to legitimacy, yet they exist (either physically or virtually) and provide another interesting, albeit fantastical, mix in the study of small territories. These non-nations' lack of legitimacy, their relative insignificance on the global stage, and the fact that most of them were founded as hobby endeavors or whimsical efforts to gain attention, have resulted in a lack of attention by serious scholars, yet they do exist, and many of them have become tourist attractions in their own right (Timothy, 2003). Table 10.2 lists several of Europe's most prominent micronations.

One of the most successful micronations has been the Principality of Hutt River in Australia, which came about after a wheat quota dispute between a farmer and the Australian government pitted one against the other. The farmer, Leonard Casley, declared independence on April 21, 1970 (de Castro and Kober, 2018; Ryan *et al.*, 2006). Aside from agriculture, the 'country's' main

Table 10.2. Selection of self-declared micronations in Europe

Micronation	Location	Declared	Origins
Saugeais	France	1947	Came about as a result of a joke by a local resident toward a government official visiting from another region. The name and location remained in local lore and is a salient tourist attraction.
Liberland	Serbia & Croatia	2015	Liberland was formed on land that is disputed between Serbia and Croatia near the Danube River. As neither country claims the land, Vít Jedličk declared the 7 km^2 area the Free Republic of Liberland in 2015.
Užupis	Lithuania	1998	Located in an arts district of Vilnius, Užupis declared independence from Lithuania and Vilnius to be a home for an artisan community where artists could mingle and share their talents.
Austenasia	UK	2008	Self-declared sovereign state in London but including properties throughout the UK. Formed when the Austen family declared independence for their home in Sutton, London.
Sealand	UK	1967	A disused World War II [also Second World War] anti-aircraft sea fort/platform 12 km off the east coast of England. Claimed by the Bates family and declared independence from the UK in 1967.
Pontinha	Portugal	2007	In 2000, Renato de Barros bought a small islet off the coast of the Portuguese island of Madeira and declared its independence from Portugal in 2007.
Jamptland	Sweden	1963	The area was self-governing in the Medieval period, and owing to the local authority consolidating rural areas, the micronation was established in 1963.
Ladonia	Sweden	1996	Independence declared by two artists in a dispute with the local municipality regarding how their public art was categorized for permitting purposes.
Seborga	Italy	1963	According to its founders, historical documents suggest that the medieval principality was left out of the treaty of 1861 that unified Italy and therefore remains a sovereign principality.

Source: based on information in Middleton (2015), Ryan *et al.* (2006) and Timothy (2003)

growth sector became tourism; tourists traveled great distances to see this isolated non-nation in Western Australia, with its official border crossing and passport inspections, national flag, parliament, princely family, museum, church, stamps and post office. Prince Leonard fought for legitimacy through the Australia courts but never received the political and legal recognition he desired. The self-styled monarch died in 2019, passing the throne to his son, Prince Graeme. Since the 1970s, the Casleys developed a reasonable tourism industry based on their declaration of independence and the stately artifacts and policies they developed. However, as of January 31, 2020, the 'principality' appears to have closed its doors to tourists because of economic insolvency (Juanola, 2020).

Most other micronations have not achieved such a high level of notoriety and tourist interest, nor have they fought as hard for legal legitimacy. Nevertheless, there are dozens of these non-nations throughout the world that have either unilaterally declared their sovereignty in an effort to pursue a hobby, as an insincere joke or an overtly flippant cultural endeavor, to protest perceived government mistreatment, or in an effort to draw attention to a specific cause (Dunford, 2009). Most of them have a specifically delineated territory; this has included a bedroom or individual home, a garden or backyard, an abandoned building, an unpopulated island or plot of land, an urban neighborhood, a village, or a virtual meeting space.

National governments are unaware of their existence, except for a few cases (e.g. Hutt River, Pontinha and Seborga) where leaders have legitimately attempted to negotiate their independence through legal means. Timothy (2003, pp. 93–94) suggests that the level of legitimacy among these 'fake states' can be seen from four perspectives. First are those that have received international diplomatic recognition as a sovereign entity but not a state, such as the Sovereign Military Order of Malta discussed in Chapter 9. The second type are places that may have some legitimate historical claim to independence and statehood but whose claims have not been ratified by other states, especially the one that enfolds the faux nation (e.g. Italy's view of Seborga). Third are those non-nations that may not have any historical or legal grounds for legitimacy but have nonetheless attempted to seek legal recognition (e.g. Sealand). Finally, the most plentiful type are those that were established in jest as a practical joke or a humorous exploit, an effort to establish a hobby group or social club, a means of realizing a personal dream or fantasy, or simply an effort to create a virtual space for like-minded people on the internet (Serzhanova, 2014).

The majority of these fictional countries have several things in common. First, they are declared by people carrying out their personal fantasies or dreams. Second, they make every effort to produce symbols of sovereign states, even if for touristic or entertainment purposes. These include coins and banknotes, postage stamps and post offices, passports and border checkpoints, constitutions, coats of arms, flags, national anthems, car license plates, monarchies or prime ministers, and complementary citizenship for interested participants. Third, they all lack legal, geographical and political legitimacy. A few of them have succeeded in establishing 'diplomatic missions' with other make-believe non-nations and setting up faux 'embassies' abroad, but this is part of the humorousness that drives their existence. Fourth, people can buy souvenirs and nationalistic paraphernalia from these far-flung localities online without actually visiting. Online sales of citizenship and passports, coins and stamps, bumper stickers and t-shirts, recordings of national anthems, flags, and many other symbols that 'authenticate the inauthentic' are an important income source for some micronations and provide virtual visits for people who are unable to visit in person. Finally, they are all curiosities that tourists, travel writers and the media in general latch on to for their novelty value. Many non-nations have become salient regional destinations, as the Hutt River example above illustrates. In Europe, several of them have also become a local focus of tourism and provide an additional competitive advantage for the destinations where they are located.

Užupis

On April Fool's Day, 1997, a group of artists in Vilnius, Lithuania, declared their historic neighborhood, Užupis, an independent country. This was done in jest, figuratively as a way of helping promote an artisan lifestyle in this area of the city (Middleton, 2015). Slowly word about the 'independent' artist colony spread, and today it has become an interesting tourist spectacle in Vilnius. Largely because of tourism, the Užupis area has become one of the most visited and gentrified parts of the city, which has priced out many of the artists whom it originally aimed to attract. Every April 1, Užupis celebrates its 'independence', offering passport stamps, opportunities to spend the local currency, and partake of free-flowing beer from the fountain in the main square (Rhone, 2018).

Užupis has a self-proclaimed president, constitution and currency, and although it is not a legitimate state, the novelty of the places has, according to Rhone (2018), become a point of pride throughout Lithuania and one of Vilnius' most notable attractions. The importance of Užupis in Vilnius' tourism landscape is notable in the city's marketing efforts. Even the local destination management organization, Go Vilnius, proudly encourages tourists to visit the unique micronation:

> Užupis is the smallest district in Vilnius and is separated from the Old Town by the Vilnelė River. Formerly a troublesome place in the early 1990s, Užupis has grown into a beautiful

> neighbourhood of artists, intellectuals and entrepreneurs. On 1 April 1997, residents of Užupis decided to somewhat sarcastically declare independence. An independent republic has to have money, power, an anthem and constitution, and above all, a free spirit. You will find all of this in Užupis. Make your way under the Užupis Bridge to have a go on the Swing of Destiny and make sure to say hello to the Mermaid of Užupis while you're down there. (Go Vilnius, 2020, n.p.)

Saugeais

The Republic of Saugeais is located in eastern France against the border of Switzerland. It encompasses several municipalities and has a fairly large area of 128 km² and a population of around 5000. This micronation began with a joke in 1947 when a government official from a different region arrived in the Saugeais valley for lunch and the restaurateur jokingly asked if he had a permit to enter the territory of Le Saugeais. The visiting official enjoyed the joke, and after inquiring about the history of the area, suggested that Saugeais seemed like its own country and invited the restaurant owner to be the president of the *La République du Saugeais* (Eyres, 2020). This started a unique tradition in the area that has lasted until today, being nourished by tourists' curiosity and local events that promote the region.

In the years that followed, the micronation developed its own coat of arms, flag, anthem, coinage, president and prime minister, and tourists can stop by the gated 'border checkpoint' for a visa and passport inspection. The pretend republic and its various state paraphernalia are the main attractions in the area, although the restored abbey and small Saugeais museum are important regional cultural assets. The valley is also known for its smoked hams and sausages, which form the foundations of a growing gastronomic tourism sector (Eyres, 2020). Although this fake state began as a humorous punchline, it has become part of the local lore and a project in which some in the community, especially the founding Pourchet family, have taken great pride and developed a considerable tourism industry (Ryan *et al.*, 2006).

Seborga

The Principality of Seborga is probably the European micronation with the most reasonable historical and legal foundation for its sovereignty claims (Timothy, 2003). Its claims stretch back to the year 1079, when it became a principality within the Holy Roman Empire. From the eleventh century until the nineteenth century, when other principalities were mentioned by name in land transfers and territorial consolidations, the Principality of Seborga was never mentioned. In 1729, the royal House of Savoy is said to have purchased Seborga, but the Savoys never registered the transaction. As well, unlike other monarchies and republics on the Italian Peninsula, Seborga remained outside the treaty that unified the Kingdom of Italy in 1861 (Calvini, 1992; Timothy, 2003). Again in 1946, when Italy became a republic, Seborga was not mentioned in any of the agreements. Based on his archival research, Giorgio Carbone, a local history enthusiast, argued in 1963 that all of these historical omissions demonstrate that the Republic of Seborga does not, in fact, belong to Italy. A local referendum took place in 1995 on the question of sovereignty, and the majority of the village's inhabitants agreed and reconfirmed Giorgio as Prince of Seborga (Middleton, 2015). The monarch died in 2009 and a new prince was elected in 2010 and re-elected in 2017.

The community has petitioned Italy for recognition, but Italy and most observers see the situation as a fantasy game (Ryan *et al.*, 2006) and refuse to accept the ostensible evidence discovered by Mr Carbone. The village of Seborga today functions as an integral part of the Italian state. It is 14 km² with a population of about 300 (Fig. 10.10). It is located in a scenic mountain area in far north-western Italy in the Liguria region near the border of France. In 1995, the principality began minting its own coinage in the local currency, the luigino. The principality's postage stamps are rare collectors items and can be purchased along with coins and a wide range of other souvenirs, although only Italian stamps can be used to send postcards, letters and packages from the local post office. Among the non-nation's most important attractions is its 'official' border crossing with Italy (Fig. 10.11).

Fig. 10.10. The central piazza of Seborga with government offices and souvenir shops

Fig. 10.11. This staged border checkpoint is an important attraction in this micronation

The Liguria region's marketing agencies include promotional materials—brochures and websites—about Seborga as a sovereign principality in a tongue-in-cheek sort of way because it is good business for the region's tourism sector (Regione Liguria, 2017). One regional brochure touts the idea that 'Seborga is the only village in Italy to have a mayor and also a prince' (Regione Liguria, 2017, p. 2). If Seborga's efforts ever pay off and Italy legally recognizes the village's claims to sovereignty, the principality will be the third microstate to perforate Italy.

Conclusion

The de facto states and micronations highlighted in this chapter face many challenges on the global stage. Primary among these is a lack of recognition by all except their patron state and a small handful of other unrecognized states and territories. This renders them powerless in terms of trade in general terms, but tourism is a curious phenomenon. There will always be intrepid travelers who want to visit a place just because no one else does, or because it presents a set of unique challenges. As such, as tourists continue to seek off-the-beaten-path destinations, many of them are looking to de facto states to satisfy their geopolitical curiosities. From the perspective of quasi-states, 'We are far from mass tourism, but statistics are showing an increasing interest in unrecognized territories and an increase of tourism in Georgia and Ukraine is often connected to an increase of the numbers of tourists going to Abkhazia and Transnistria, too' (Lambert, 2018, p. 119).

The sorts of tourism associated with de facto states in Europe resemble various forms of dark tourism, or even war zone tourism. There is a fascination associated with stepping foot inside unrecognized polities, participating in their illicit economies and politics, spending local currencies, and visiting the 'forbidden other'. For de facto states, this is their magical tourism asset, although many, such as Artsakh and Abkhazia, have legitimately one-of-a-kind natural areas, beaches and heritage sites that could attract tourists even if they were not in a state of legal limbo.

Reunification with the legal home state and increased collaboration or improved relations could cause the competitive advantage of the secessionist dispute to disappear which, while good for political relations, might not be so good for tourism. Nonetheless, current research on peace and tourism suggests that benevolent relations are more conducive overall to the growth of tourism (Gelbman, 2010; Jacobson *et al.*, 2015). Jacobson *et al.* (2015) argue that, despite Northern Cyprus' unique appeal, and likely the same in other de facto states, reunification would be a net gain for tourism on both sides of the border.

11

Vulnerability, Dependency and Touristic Appeal: The Costs and Benefits of Smallness and Marginality

Introduction

As denoted throughout the previous pages of this book, microstates, small territories, and other anomalous polities have received considerable academic attention, but most of this attention is repetitious and focuses on the interconnected concepts of vulnerability and dependency (Bartmann, 2002; Catudal, 1975, 1979; Vinokurov, 2007). Although all countries and regions of the world, including small places, face certain vulnerabilities, such as security threats, climate change and natural disasters, some vulnerabilities are unique to small political entities. In the socio-economic and political context of tourism, smallness, isolation and differentness sell well, but they also raise challenges to tourism and other means of development. This chapter discusses the costs and benefits of tourism associated with smallness and marginality, under the conceptual umbrella of vulnerabilities of size, overdependence, lack of power, lack of recognition, and the appeal of 'otherness'.

Vulnerabilities of Size

Small size, small population, access problems, and inequitable power relations, are what define tourism vulnerabilities in the microstates and anomalous polities of Europe.

Small geography

The most commonly stated challenge for microstates, exclaves and small territories is their diminutive size (Baldacchino and Wivel, 2020; Harrison, 1996, 2004; Thorhallsson, 2018; Veenendaal and Wolf, 2016). This necessarily limits the diversity of natural resources available for development (Armstrong and Read, 1994; Carlsen, 2017; Connell, 1991; Conway, 1998). Thus, most microstates are overdependent on one or only a few economic activities (Baldacchino, 1993), which means that most required consumer items must be imported (Briguglio, 1995, 2016a).

In terms of tourism, small physical territory is an obvious obstacle to achieving similar levels of tourism development as in larger states and territories. Smallness affects tourism negatively in two primary ways. First, as with natural resources in general for economic development, extreme smallness also signifies a scarcity of tourism assets. Limited land area restricts the range of resources that can be used for tourism (Liu and Jenkins, 1996; Timothy, 1996, 2001a, 2001b). Tourism resources and products in larger countries are usually more diverse, with a combination of activities and localities targeting heritage and cultural tourists, sport tourists, religious tourists, nature-loving tourists, and others. In the smallest polities, there is a danger that tourism might come to depend on only a few attractions and tourist types, resulting in their overuse and susceptibility to exogenous forces, such as overcrowding, as in the case of Vatican City, San Marino City, and Monaco, or resource depletion.

Likewise, small countries usually have a market disadvantage when having to compete with larger destinations that have similar products and a better geographical advantage (Timothy, 1996). Most tourists prefer a destination that has world-renowned monuments and a range of assets and activities rather than one or only a few. Very few of Europe's microstates and other unusual polities have world-class attractions, the way Europe's larger countries do. Scale is an important geographical concept in understanding this problem. What might be a 'major' attraction and extremely important symbol of nationhood in a mini state (e.g. the ruins of Obere Burg and Untere Burg in Liechtenstein and the former sixteenth-century parliament building, Casa de la Vall, in Andorra) might not even make the cut for a brochure or website in many larger countries. Yet, dichotomously, this relatively atomic scale also adds considerable charm to small places.

A limited attraction base, together with easy access, will induce a greater flow of less profitable same-day trips rather than more lucrative overnight stays, which is the case in Europe's continental micro-polities. Jenner and Smith (1993, p. 69) insist that 'It is difficult...to encourage tourists to stay long in a country you can drive all the way through in anything from a couple of hours to as little as a few minutes'. Because of its globally celebrated art and architecture, Vatican City may be the exception to this, where many of its visitors spend all day in the country, but without lodging facilities they cannot spend the night, even if they want to.

Extraordinary small size has implications for domestic tourism as well. Even small, but larger, countries such as Singapore and Luxembourg, pay little attention to domestic tourism. Because of their small size, 'domestic tourism' in microstates may need to be counted differently to how the World Tourism Organization normally counts tourism (overnight stays). Most microstates do not concern themselves with measurable 'domestic tourism' (Timothy, 2020). If they did, it would need to be re-defined for small polities and would look something like recreational outings closer to home, which Canavan (2012) calls 'micro domestic tourism'. Timothy (2020, p. 70) argues that 'length of stay (overnight) may not be as applicable a variable in microstates as it is in larger countries'.

The second way in which small land area affects tourism negatively is that a polity's compactness inhibits its ability to expand tourism assets and infrastructure physically. The compact size of microstates, enclaves and several other entities physically restricts the planning and development of tourism (Timothy, 1996), as in Andorra, where previous attempts to build an airport or develop large attractions have failed because there were no locations large enough to accommodation them (Timothy, 2001a). Monaco has long had big plans for tourism and other means of development, but it lacks the physical space for the growth it needs to achieve its goals. Although hampered considerably by its tiny geography, Monaco is the only continental microstate or territory discussed in this book that can extend its geography physically and legally, because it is not landlocked (d'Hauteserre, 2001). The principality shares the Mediterranean *Côte d'Azur* with France, its sovereign territorial waters extending 12 nautical miles into the sea. Since the late nineteenth century, the country has extended its physical territory considerably, with reclaimed land now comprising approximately one quarter of the country's total land area. As explained in Chapter 5, the current reclamation project will add more surface area for high-end apartment buildings, businesses and retail space. Although in theory Monaco could continue to extend its territory further, there are significant environmental concerns and EU stipulations to consider, and encroaching upon France's territorial waters cannot occur.

Physical compactness sometimes means that small territories and microstates are also encumbered by physiographic challenges (Burnett and Uysal, 1991; Reichart, 1988). Andorra's location in the Pyrenees means that it has had to contend with physical geography for centuries (Taillefer, 1991). Because of its size limitations, it has no territory outside the mountains, although it has made significant strides to overcome its physical limitations by building a strong network of tunnels and viaducts, as well as sponsoring the construction of an airport at La Seu d'Urgell, Spain. Monaco also faces topographic challenges on the coastal edge of the Maritime Alps. The principality's steep topographic gradient from sea level to 160 meters has been a significant impediment for transportation, walking and urban development, although the country has overcome this problem by installing a network of lifts and escalators throughout its small area. The majority of Liechtenstein's national territory is mountainous, but it has been able to develop a large network of roads and tunnels, and most development has occurred in the Rhine River valley.

Small demographics

Small populations are another defining characteristic of microstates. Small domestic markets in microstates and territories translates into diseconomies of scale. Diseconomy of scale means 'The unit cost of local output of...goods and services will be relatively high because the level of domestic demand lies below the minimum efficient scale' (Armstrong and Read, 1998, p. 567). Thus, it is unfeasible to produce consumer items and even many services at a cost-effective rate to satisfy the needs of a small population. Therefore, many services are outsourced to neighboring countries. This decreases domestic competition since small populations cannot support more than a few firms selling the same product, which sometimes results in the development of monopolistic enterprises (Briguglio, 1995, 2016b). Small domestic markets also increase small states' dependence on foreign imports and the need to compete internationally for export markets against larger producers, limiting a microstate's export ability and market (Allegrezza *et al.*, 2019; Armstrong and Read, 1994; Briguglio, 1995, 2016a). Nonetheless, microstates are overly dependent on exports, as a means of earning foreign exchange to pay for their requisite imports (Briguglio, 2016b). All of Europe's microstates and most of the other special polities have chosen tourism as either their main export sector or one of their most important.

Small populations and miniscule domestic markets may repel potential investors in many social or development programs. Foreign direct investment (FDI) usually seeks large economies that will provide generous returns on an investment (Allegrezza *et al.*, 2019). There is clearly an element of danger in investing in unrecognized states, which tend to be unstable and volatile. As well, some of them (e.g. Turkish Republic of Northern Cyprus) are under international boycotts, which proscribe FDIs in the first place. Oxymoronically, however, many foreign tourism investors have sought monetary destinations whose sustainability legislation is weak or ineffective in order to save money on development that does not adhere to environmental standards. Some local officials, Dinica (2012) suggests, create these environmentally relaxed locations as a way of maximizing foreign investments. Such a perspective does not hold true in the European microstates, which are all well legislated in their own national laws, as well as in how they conform to EU environmental regulations. Such standards may not necessarily apply in de facto states..

In many cases, small populations also limit the pool of skilled human resources to work in various sectors, including tourism (Baldacchino, 1993; Carlsen, 2017; Carlsen and Butler, 2011; Katnic and Boskovic, 2019; Milne, 1992). Many skilled workers migrate from microstates and other polities because of a lack of adequate jobs at home and better opportunities for career development abroad (Connell, 1993; Connell and Conway, 2000). This is less problematic in Europe's smallest states where ordinary tourism and hospitality positions are often filled by immigrant workers or laborers who cross the border daily and where management-level skills are more abundant.

Access issues and the vulnerabilities of location

Access may refer either to physical access or market access, with market access referring to cost, time and perception rather than physical reachability. Traditionally, from a legal and geopolitical perspective, the right of (physical) access has been a paramount concern in many of the European exclaves and anomalous polities. Although friendly relations between home and host state have helped alleviate accessibility problems (Catudal, 1979; Vinokurov, 2007), concerns remain even today that enclaves could be held 'hostage to current relations between the two states in question' (Forysinski, 2012, p. 27).

Law enforcement has been a particular concern, as most sovereignty laws do not allow armed or on-duty police or military to cross another state's territory. In a few enclave cases, bilateral agreements have allowed home state police to pass through a host state to reach the exclave. Likewise, the freedom of passage for exclave residents to reach the homeland, as well as tourists traveling between the home and host state, has been a historical concern (Berger, 2010; Forysinski, 2012). However, today such challenges do not exist in the European exclave situations, as essentially all borders are open, with the exception of the new checkpoint established between Switzerland and Campione d'Italy in 2020.

In West Berlin's former outliers enclaved in East Berlin until 1990, neutral corridors were established to allow West Berlin's police and citizens to get to the outliers in the east (Catudal, 1971; Lundén, 2012; Robinson, 1953). Similarly, to alleviate concerns over citizen, government official, and police access to Llívia, France built a neutral road, which could only be used by people transiting French soil between the Spanish mainland and the enclave (Catudal, 1979; Vinokurov, 2007). Today, the neutrality of that road is less relevant, because both countries are part of Schengen and the common market. Through a bilateral agreement, French and Spanish police are today permitted to cross into the territory of the other state as necessary up to 20 kilometers on official business.

Although traditionally, enclaves, dependencies, and territories in Europe faced isolation from domestic markets given their separateness from the state homeland (Timothy, 1996), this is of little concern today as the Faroe Islands, Isle of Man and Channel Islands, are more accessible now than ever before through frequent flights and ferry services.

Physical isolation raises concerns both about physical access and market access (Ankre and Nilsson, 2016; Connell, 2007; Lockhart *et al.*, 2002; Ryan, 2001). A lack of an airport or distances far from airports (e.g. Andorra, Transnistria, Artsakh) can keep people from visiting, who might be put off by high transportation costs or time constraints (Briguglio 1995; Carlsen and Butler, 2011; Timothy, 2001a). Some of these people will choose destinations that are more convenient to reach and which have greater tourism offerings.

Overdependence

The second category of challenges most discussed in the literature is overdependence, which in its own right is a symptom of small size, limited demography and peripherality, and itself seething of political vulnerabilities. Historically speaking, dependency theory holds that core states were enriched by exploiting the resources of their colonial outliers, leaving the colonies with little foundation for their own growth and development. The implication of core-periphery relationships was that the colony remained underdeveloped and dependent upon its European metropole for its survival. This meant that some colonies were overly dependent on a single trading partner, even after independence. Although dependency theory has been heavily critiqued (e.g. Gosh, 2019), some of its basic elements are evident in the patron-client relationships in Europe.

This section addresses the distorted relationships between smallness, marginality, patronage and clientelism that result in conditions of 'overdependence'. For our purposes, there are two primary ways of understanding overdependence: overdependence on tourism (Britton,

1987; Lea, 2006) and overdependence on the exogenous 'other' (Maass, 2017; Misra, 2004).

Overdependence on tourism

Tourism has historically been targeted as the inevitable contingency economy for physically and demographically small and isolated states and territories where other sectors are less feasible (Cuevas Contreras *et al.*, 2016; de Albuquerque and McElroy, 1992; Fagence, 1997, 1999; Hall, 1994a, 1994b; Liu and Jenkins, 1996; Milne, 1990; Pratt, 2015; Sharpley, 2012). Overdependence on tourism has been problematic for many small islands and microstates at least for two reasons.

First, lacking a diversity of fiscal sectors means that local economies will necessarily suffer from natural disasters, global economic downturns, and worldwide pandemics. Tourism is one of the most inelastic and volatile industries on the planet. Even whispers of political discontent, disease, or economic slumps will send arrivals tumbling. When a small economy is based entirely or even primarily on tourism, it is dangerously susceptible to external forces it cannot control (Britton, 1982, 1987). While generally not as obvious or intense as small island developing states (SIDS), many of the small territories and microstates in Europe are heavily dependent on tourism and are also affected by forces beyond their ability to manage. Although tourism tends to be less dominant in Europe's pseudo-states/de facto states, most of them are in a constant state of conflict, so whenever tensions erupt, the tourism sector declines dramatically in areas already suffering from peripherality.

Secondly, tourism-oriented small states and territories are often at the mercy of outside industry and policy decision makers (Baldacchino, 1994, 2019; Timothy and Ioannides, 2002; Wilkinson, 1989). In Baldacchino's (2019, p. 41) words, 'the smaller the state or territory, the greater the likelihood that its domestic, internal affairs will be dominated, responsive to and driven by exogenous factors'. Airlines, cruise companies, tour operators and hotel corporations based elsewhere regularly make decisions that affect domestic markets, frequently without input from the destinations and populations most affected by those decisions. This resembles the notion of 'neocolonialism', wherein destinations suffer the burdens of tourism (i.e. negative sociocultural, economic and ecological impacts) without benefiting much from the industries that exploit them, similar to conditions during colonial times. Relatedly, many less-affluent mini states depend on outside investments, such as foreign (expatriate) ownership of tourism services (Milne, 1992). This means that many dividends will leak out of the local economy and return to the bank accounts of foreign entrepreneurs. Fortunately, this is not as relevant in Europe as it is in the Caribbean and Pacific microstates, but it can be a problem in some of the smaller territories and de facto states. As previously noted, small economies may also mean that foreign investors are less likely to invest. However, tourism seems to be a special case. Where the industry is thriving (e.g. microstates), outside agents are more likely to invest in a lucrative tourism sector.

Overdependence on the exogenous other

Many small and anomalous polities demonstrate a core-periphery relationship (Armstrong and Read, 1998; Hepburn and Baldacchino, 206; Maass, 2017; Sharpley, 2001; Vital, 2006; Weaver, 2017), wherein the periphery (the outlier) relies on the core (the homeland or patron state) not only for support but also even for 'permission' or approval to enact tourism policies or legislation. Observers have long argued that this dependency relationship is a barrier to development, although Sharpley (2003) contends that it does not have to be. The exclaves and pene-exclaves of Europe maintain a pragmatic dependency relationship with their enclaving states. As Vinokurov (2007, p. 205) correctly points out, in matters of everyday life and tourism, enclaves are 'highly dependent on mainland-surrounding state relations for peace, stability, and well-being'. They are often also dependent on foreign institutions centered in the protector state, such as legal services, universities/trade schools and banks (Connell, 1991; Wolf *et al.*, 2018).

These patron-client/core-periphery and dependency relationships frequently manifest in an over-reliance on a single market, for example Russians in Abkhazia, and Swiss in Liechtenstein and Campione d'Italia. Unrecognized states demonstrate this single-market over-reliance very well, especially when they are under international embargos. Northern Cyprus, which is still under international trade and travel sanctions (even though tourists are free to visit from the south), for many years could only receive tourists from Turkey (Seddighi *et al.*, 2002; Yasarata *et al.*, 2010). Even today, only flights from Turkey can arrive at its airport. Likewise, since the early 1990s, Abkhazia has been under various blockades by Georgia, Turkey and the Commonwealth of Independent States. It has survived economically largely through Russian inbound tourism (Popescu, 2006a). This pattern of reliance on a single market resembles the colony-metropolitan relationship still apparent in the Caribbean and Pacific (Conway, 1998; Hepburn and Baldacchino, 2016) with regard to colonial travel, especially among current and former French colonies and metropolitan France. This is less obvious in the British Caribbean where the primary UK market for British territories and former colonies generally comes second to Americans and Canadians owing to geographical proximity.

Related to the previous two points is an issue that befalls many microstates, though less so in Europe than elsewhere: the growth of unhealthy and unbalanced bilateral relationships between a microstate and a larger power broker state, often underscored by intimidation and economic strong-arming. In the 1980s, the Soviet Union convinced Kiribati and Vanuatu to let it rent their fishing rights (Connell, 1991), whereas other small states in the region chose to lease their rights and align more closely to the European Community/European Union. Today, part of the political discord currently between the People's Republic of China (China) and the Republic of China (Taiwan) is playing out on the Pacific battlefield as the two Chinas seek favor of a handful of tiny SIDS, forcing them to choose sides (Stringer, 2006). For example, because Palau has chosen to align more closely with Taiwan, China recently severed all tourism ties with Palau, declaring the microstate an 'illegal destination', cancelling all flights and prohibiting holiday bookings (Timothy, in press). In September 2019, two of Taiwan's supporters, Kiribati and the Solomon Islands, ended diplomatic relations with Taiwan, instead transferring their diplomatic preferences, trade activities, and tourism support to China. In response to this action, several other microstates, including the Marshall Islands, Palau, Tuvalu and Nauru confirmed their continued support for Taiwan, pitting neighbor against neighbor in this diplomatic standoff (Timothy, in press). Although this is more common now in the Pacific, some of the patron-client relationships involving Europe's proto-states resemble these disempowering and conflicting relationships described above. Sometimes, coercion and promises of financial security are used as tools to cement client-patron relationships, making reconciliation with the motherland almost impossible and in some cases resulting in armed conflict (e.g. Luhansk, Donetsk and South Ossetia) (Metreveli and Timothy, 2010).

Lack of Power

Power, self-determination and autonomy are similar concepts that denote a certain level of authority to exercise one's will and administer one's own affairs. Of all the special units described in this book, the microstates are the most empowered, given that they are sovereign polities. Yet, even these are not entirely emancipated from their larger overseers, as for example with international security and diplomacy (Goetschel, 1998). The autonomous regions and territories of Europe are not sovereign states, but they have more internal authority than other areas of the state to which they belong. Exclaves are essentially powerless, functioning at the mercy of the motherland and the host state, and the de facto countries are effectively puppet states controlled by a master puppeteer with tight strings.

The powerlessness of many European geopolitical entities historically has been most palpable in their susceptibility to the mercy of a patron state or enclaving state. The Turkish Republic of Northern Cyprus, for instance, is entirely reliant on its enabler state, Turkey, for economic sustenance and tourism. Turkey dictates much of what the TRNC can do in tourism actions and policy-making (Bahcheli and Noel, 2010). Another poignant example is France's unilateral blockade of the Monégasque border in 1962–63 in retaliation for the principality exercising its sovereign rights to allow French citizens the right of abode without paying French taxes (Bézias, 2007; Catudal, 1975). Only after France's political 'persuasion' did the microstate change its tax laws regarding French expatriates and their fiscal responsibilities, enabling the border blockade to end peacefully. Similarly, earlier in the twentieth century, San Marino planned to build a casino to earn additional revenue just as Monaco had done, but Italy disagreed with the notion and blockaded the country until it relented (Catudal, 1975).

Outliers, autonomous regions, and overseas territories are fundamentally powerless to deal with certain international issues that affect them directly but must be dealt with at the level of the sovereign state. Thus, decisions made at the state level affect them, even if the decisions are counterintuitive or counter to the wishes of the local population (Amoamo, 2018). A good recent example is Gibraltar and Brexit. Even though 96% of Gibraltarians voted to remain in the EU, not being a sovereign state, the territory was required to leave the EU with the UK in January 2020, raising considerable fears of a Spanish border lockdown, reminiscent of the 1969–1985 closure. Gilbraltar is exceptionally reliant upon Spanish day workers, cross-border tourists, EU development funds, and business access to the common market. Gibraltar stands to lose a lot from this geopolitical showdown (Mut Bosque, 2017).

Compared to microstates, exclaves have an even higher sense of powerlessness, as they exist in conditions of quadruple peripherality. First, they are by definition in the borderlands, on the national margins, which have traditionally been regarded by the state as tangential areas both politically and fiscally. Second, they are physically separated from the sovereign state to which they belong. Third, they are usually very small in area and population, so that they cannot support certain services, such as schools and health care facilities. Finally, because they are separated from their motherlands and usually part of a municipality or county within the home state, they lack decision making power over even mundane concerns.

This can be illustrated by a few examples. On a couple of occasions, France or its Pyrénées-Orientales department has had conflicts with Llívia. The enclave receives its water from the French. The area of Pyrénées-Orientales receives its water supply ahead of Llívia, and several times in the past, it has cut off the enclave's water supply for various reasons (Eggers, 2019). In the 1970s and 1980s other disputes erupted between Llívia and French officials, when the local French authorities erected unwelcomed stop signs to allow the right-of-way for French drivers crossing the neutral road. This led to political bantering, as well as vandalism and destruction of the signs and an eventual short-term blockade of the exclave by French authorities. Events such as this one in 'normal' situations would be considered local and somewhat trivial. Yet, in the context of Europe's small polities, this turned out to be an international diplomatic crisis that had to be mitigated at the state level.

In the Baarles, outbreaks of foot-and-mouth disease led Dutch officials to restrict the transport of Belgian livestock over Dutch territory. This had major implications in the early 2000s for farmers in Baarle-Hertog, who were entirely surrounded by the Netherlands. All farm animal transport was restricted into and out of the Belgian exclaves, leaving several of the Belgian farmers unable to sell their products in Belgium proper just a few kilometers away. Many accounts such as these in Monaco, Llívia and Baarle-Hertog have happened at enclave and microstate borders throughout history, most without any legal recourse for the smaller polity (Butler and Morris, 2017).

Efforts by Moldova to block access to Transnistria and attempts to block access to Artsakh by Azerbaijan have led to armed skirmishes. Small territories and political outliers have often been the subject of international conflict, over which they are powerless to intervene. Morocco continues to contest Spain's long-held sovereignty over the Spanish Plazas de Soberanía on the Mediterranean, and Spain has challenged the UK's sovereignty over Gibraltar since the Franco dictatorship began in 1936 (Gold, 2005). This has clear consequences for tourism's peaceful growth in Gibraltar and the Plazas de Soberanía. This was especially evident when Spain unilaterally closed its border with

Gibraltar in 1969 and did not re-open it fully until 1985 (Berger, 2010; Seekings, 1993).

According to Vinokurov's (2007, p. 202) synopsis and analysis, Catudal's (1979) main 'exclave problem hypothesis' is that the surrounding state wants to absorb the enclave, while the legal home state wants to ensure standard administration over the outlier just as it has over the rest of the country, and finally that these two interests are incompatible. That might have been the case some decades ago, but the European internal market and the Schengen mobility zone has reduced any desires to take over enclaves in Switzerland, the Netherlands, Belgium, Italy, Germany or France. Even today, however, tourism-oriented exclaves still have to navigate the imbalances of power between the home and host countries (Timothy, 1996). Mehmet and Tahiroglu (2003) question whether microstates, being small and 'powerless', automatically lag behind normative states in the process of globalization. This does not appear to be the case in Europe, where they seem to be doing well on their own as well as in relation to their supranationalist ties.

Their apparent lack of power means many microstates and small territories are exposed to shocks in the international economic and global trade environment (Armstrong and Read, 1998; Briguglio, 2016a; Briguglio and Vella, 2019; Katnic and Boskovic, 2019). In Europe, this has caused them to hasten their effort to join, or affiliate with, larger trading blocs for their own economic protection and very existence. Although all states, large and small, are at the mercy of worldwide markets and economic conditions, microstates typically have to work harder than larger states to be accepted into the global community of independent countries and to survive in the global marketplace (Brugiglio, 1995; Vital, 2006). For centuries, microstates have been annexed, overrun or absorbed because they were defenseless against larger countries (Diggines, 1985). Given the current global political climate and the increasing integration of the microstates with the European Union, it is unlikely that this would occur today, yet there are still mechanisms within the patron-client relationship that can disempower small states and territories for the benefit of the puppeteer.

Lack of Recognition

Although much improved over the past 30 years, in some cases large states do not take the microstates or exclaves seriously. Microstates are frequently overlooked in the international political and diplomatic system (Stringer, 2006). This became very clear in 1919–1920, when Liechtenstein, Monaco and San Marino attempted to join the League of Nations, the forerunner of the United Nations. In 1920, the League rejected Liechtenstein's application to join, ostensibly because it had delegated too much of its sovereignty to Switzerland, including defense, to be considered a fully independent country in the peerage of sovereign states (Dumieński, 2014), although it is more likely that this was the League's way of politicizing the microstate's small size and the organization's failure to acknowledge Liechtenstein as a sovereign state. Liechtenstein later (1949) applied to be a signatory to the International Court of Justice, but the Soviet delegation complained that the microstate was not 'a free and independent state' (Bartmann, 2012, p. 542), so its application was delayed. Large states have historically not valued the microstates. The concept of large is normal and preferable, whereas small is abnormal if considered at all, or 'at best petty and Lilliputian' (Baldacchino, 1993, p. 30).

Taking a clue from Liechtenstein's 1920 failure, Monaco and San Marino withdrew their own applications to join the League (Bartmann, 2012). According to Bartmann (2012, p. 537), the main problem was their small size, layered with questions about the validity of their sovereignty. The historical terminologies used by the community of states reveal a great deal about how they were regarded: 'statelets', 'quasi-states', 'ceremonial states', 'accidents of history', 'political oddities' and 'confetti states'. Based on European history, the small states were seen as 'fragile creatures in the rough sea of international relations. They were internally well suited for democratic regimes but externally helpless and constantly threatened by extinction' (Goetschel, 1998, p. 113).

Many League of Nations members could not fathom the idea of allowing the tiniest spots on the map of Europe to be admitted as an equal. The UK did not want Monaco to join because it was so enjoined with France that the British felt it would be akin to giving two votes to France (Eccardt, 2005, p. 23). At the 1920 formation of the League, Vatican City had not yet been created, but the Holy See was a sovereign entity. However, because it had no national territory, it did not meet the requirements of a state (Eccardt, 2005).

Prevailing prejudices against small states within the League and later in the United Nations (UN) meant that San Marino, Monaco, Liechtenstein and Vatican City could participate in the UN as observers but not full members (Catudal, 1975; Duursma, 1994) until the early 1990s when these microstates' claims to sovereignty were more palatable to the community of states (Neumann and Gstöhl, 2006). Andorra was considered only semi-sovereign until 1993, when it established a constitution that reaffirmed its sovereignty and became a parliamentary democracy. That year it was permitted to join the UN as a full member. That same year, Monaco joined the UN. San Marino joined in 1992, and Liechtenstein joined in 1990. Vatican City/Holy See has not yet applied for UN membership, but it was granted Permanent Observer State status in 1964. As regards other states' traditional views of Europe's microstates in the League of Nations and the UN, they were reluctant to admit these tiny countries because there was a:

> fear of vote dilution among the big members of the UN. Tiny Monaco would have the same power as the USA and the other big victors of the war, at least in the General Assembly...still the microstates' UN applications languished until the early 1990s. After the breakup of the Soviet Union and the admission of another huge block of members, it seemed silly to quibble about the microstates, and they finally took their place among the nations of the world—all except the Vatican, which desired to remain an observer (Eccardt, 2005, p. 23).

The most salient political challenge facing Europe's de facto states is a lack of recognition and acceptance by the world community of states (Caspersen, 2012). This has held them back politically, socially and especially economically, which has caused them, as noted previously, to depend on a single market for trade and tourism. These unrecognized states have no treaties, no trade relations (except with their patron state), and they suffer from overdependence on their patron state in trade, security and military protection, currency regulations, social insurance benefits, health care, education and tourism.

Even within their own countries, exclaves have fought for recognition and basic citizen rights. Central governments and mainland residents often neglect them. They are by definition in the borderlands where economic and social linkages are often stronger across a border than they are towards their own countries (e.g. Büsingen). They are disconnected from their home state with higher costs of access and additional border bureaucracy. They usually have only small numbers of tourists, which contributes to a prevailing air of neglect by homeland governors (Timothy, 2001a).

According to a former mayor of Jungholz, one of Austria's pene-exclaves, enclaved within Germany, 'We have to fight two formidable bureaucracies—the EU and the Austrian one, plus some offshoots of the German, although in Vienna they prefer to ignore our existence to avoid complications' (quoted in Vitaliev, 2008, p. 168). In the history of exclaves, sometimes the home country has snubbed the outliers, claiming no responsibility over these minor pieces of their own real estate. One of Büsingen's former mayors noted, 'We are in Germany, although it is easier for me to deal with the Swiss canton authorities in Schaffhausen than with our own German government in Berlin, which keeps stubbornly ignoring our problems, and at times our very existence' (quoted in Vitaliev, 2008, p. 213).

At times, the enclaving state has been equally disinterested (Vitaliev, 2008). According to one Jungholz resident, 'We are the living model of unified Europe, yet they don't want to know about us either in Vienna or in Berlin' (quoted in Vitaliev, 2008, p. 158). Yet many enclave residents feel the host state treats them more kindly than their homeland does.

At the EU level, exclaves experience an even deeper frustration. With regard to the expansion of European supranationalism, exclaves are more vulnerable than microstates. For many exclave residents, entering the EU with their mother countries is regarded as a sort of 'forced' disenclavement, which benefits them very little since their borders were already open. In fact, on the downside, EU accession has caused some of them to lose much of their former competitive advantages in tourism regarding shopping, for example. In most cases, enclaves adapt and recover from their EU disenclavement, although Campione d'Italia's return to the Italian tax regime in January 2020 will reduce its value to Swiss visitors significantly.

Vitaliev (2008) maintains that most European enclaves are examples of an ideal amalgamation of characteristics of two or more European countries which, 'like children of mixed marriages, torn between two different cultures and ways of life, they combine the traits of their mother countries with those of their host states' (p. 16). Given this advantageous feature, Vitaliev (2008) laments the European Commission's refusal to acknowledge the existence of enclaves and likens these bits of EU states to an ideal microcosmic European Union laboratory,

> where cross-cultural harmony, cross-border cooperation and 'European integration' have been happening naturally for hundreds of years. The reasons for such ostracism is simple: by their very existence, the enclaves perfectly epitomise the 'unified Europe' ideal, achieved without any assistance and/or interference from the EC, and therefore constitute a direct challenge to the very existence of arrogant and overpaid Euro-bureaucrats. 'They don't know about us, and they don't want to know', was the mantra I heard in every single enclave and semi-enclave on my route. Why? The answer was no less repetitive: 'Because—through centuries of living together—we have managed to create our own little EU—without any instructions from Brussels (Vitaliev, 2008, p. 19).

Other enclavers have expressed similar sentiments. A resident of Kleinwalsertal declared,

> enclaves are like mini-states in their own right, the only real difference being that their very existence is staunchly denied by Euro-bureaucrats...For the functionaries in Brussels, acknowledging our existence is tantamount to confessing their own irrelevance. They sit there in Brussels, Strasbourg and Luxembourg producing circulars and meaningless instructions on cross-border co-operation, which has been happening in the enclaves for ages without intervention. The legitimate question is: who needs them? (quoted in Vitaliev, 2008, p. 183).

In a tourism context, many of Europe's special polities are unrecognized in the travel marketplace. As noted previously, relatively few people in the world know about Liechtenstein, although the country is currently engaging in marketing campaigns to increase its international visibility. Thomassen (2011, p. 82) claims the same problem for the Faroe Islands, whose 'neighbours barely know of it, and hardly know anything about it'. This is even true in Denmark, where Danes 'rarely know anything about the Faroe Islands' (p. 87).

Benefits of Smallness, Marginality and Otherness

Whereas the first part of the chapter examined the vulnerabilities associated with small states and other anomalous territories, this section highlights the benefits that smallness, remoteness and otherness have in the context of tourism. Historically, economists and economic geographers believed that smallness and small economies were negative variables in the development of microstates because of their inherent diseconomies of scale and limited resources (Armstrong and Read, 1994; Carlsen, 2017; Conway, 1998; Liu and Jenkins, 1996; Robinson, 1960; Wettenhall, 2018). Although most of the world's microstates are poor and underdeveloped, it seems that small populations and small land area might be economically advantageous in the European context (Armstrong and Read, 1994; Mehmet, 2010; Mehmet and Tahiroglu, 2002). Some studies indicate that this may have something to do with their successful adoption of tourism as an economic mainstay over other industries (Algieri, 2006; Brau *et al.*, 2007; Mehmet and Tahiroglu, 2002).

The findings of a study by Armstrong *et al.* (1998) show that the most economically successful microstates and dependent territories are those that have thriving tourism industries, with banking and finance. Brau *et al.* (2007) similarly concluded that the fastest growing small states are those that are highly specialized in tourism. Small territories that have adopted tourism and other service sectors in Europe as their primary growth engines generally perform much higher than the adjacent countries surrounding them (Armstrong and Read, 1995, 2000, 2003). In the words of Vanegas and Cross (2003, p. 317), '...a growing number of studies have demonstrated that small economies surprisingly have outperformed larger economies. Thus, contrary to conventional wisdom, smallness of scale is not fatal to prosperity'.

Over twenty years ago, Baum (1997) proposed several 'fascination factors' that are inherent in small and isolated places, such as island territories, microstates and exclaves, which make them interesting as tourist destinations. These factors include small size, remoteness, overseas but not far, different but similar enough, slower pace of life and back in time, distinct culture and language, common heritage, and distinctive niche attractions. Parts of his taxonomy are used in this half of the chapter as a framework for examining many of the benefits of smallness and marginality—the variables that best characterize the anomalous polities described in this book.

Smallness

For many tourists, smallness may be a deciding factor in choosing a destination. When making a travel choice between a microstate and a larger country with similar assets and resources, many people will choose the smaller option (or both), because small places are seen as exotic or quaint (Baum, 1997; Timothy, 2001a). Visiting microstates or other small polities is particularly attractive to people from 'big places', because these offer a tangible sense of scaled-down otherness. They represent an extraordinary change from the visitors' ordinary lives. Butler (1993) submits that small islands and small countries are attractive because they provide a sense of difference and separateness. Likewise, according to Baum (1997, p. 23), 'What the visitor is experiencing is the excitement of traveling across a political border and also the opportunity to enjoy different political, infrastructural, cultural and, on occasions, linguistic experiences at the micro level'. In short, small size is a major attraction. It contributes to a place's quaintness or to the mystique associated with visiting places like Gibraltar, San Marino, Heligoland, Seborga, or Vatican City.

Some microstates eschew the notion of promoting their smallness as an attraction, perhaps because it might raise doubts in potential visitors' minds about the sufficiency of its geography and services to make a worthwhile visit. However, for the most part, in Europe this is not a concern. The microstates, many of the exclaves and territories, as well as the micronations, emphasize smallness as a crowd-drawing attribute. In its tourism promotional campaigns, Liechtenstein emphasizes its smallness as a virtue, a charm that differentiates it from its Alpine neighbors. According to one tourism brochure,

> Being small is a strength. Take a bit of time to know the 160 square kilometer (62 square mile) gem—down in the valley along the Rhine, up in the mountains near the skies. You'll be astonished: Our small size has nothing to do with numbers. Everything is here: impressive mountains and a vibrant culture, charming villages and excellent gourmet temples. Discover what it means to be small in all its diversity and familiarity. The qualities lie in the details. Whoever discovers Liechtenstein, experiences great things in a small space (Liechtenstein Tourism, 2005, p. 4).

In Liechtenstein, San Marino and Monaco (less so in Andorra), smallness is often used as a marketing point of pride—a unique selling proposition, a brand, and an asset for tourism, as they play on the notion of quaintness, smallness and otherness (Grydehøj, 2011). Witnessing a functioning state on a Lilliputian scale adds to a general sense of quaintness. A tiny parliament building, one of the world's smallest national museums, a miniature palace of justice, a more efficient postal system than the one that surrounds it (Vatican City vs. Italy), and each village having two post offices—one Spanish and one French—in Andorra are indicative of the unique or miniaturized state functions that correspond to the size of the state.

Another benefit of small size is that a rather short journey can allow visitors to see much of what a country or region has to offer. For instance, all of the main sites in the mini states, exclaves, some of the autonomous areas, overseas territories and crown dependencies can be seen in 2 days or less. Even longer visits to larger states do not allow tourists to see everything they might want to see.

Remoteness and separation (but not too far)

Remoteness can exude a sense of exoticness or adventure. For tourists who collect places for bragging rights or other reasons, even the challenges of arriving in remote destinations becomes part of the anticipated experience (Timothy, 1998a). Good examples of this include Pitcairn Island and St Helena (Amoamo, 2011). In our European situation, accessing remote and 'difficult' entities, such as South Ossetia, Artsakh, Luhansk or Svalbard can enhance the spirit of remoteness and increase the pleasure of visiting. Insularity, whether involving islands or continental outliers, is an important part of the psyche of being away from the spaces of ordinary—the separateness that people often seek (Grydehøj, 2011; Sharpley, 2001).

Different but similar enough

Socio-cultural and political differences on opposite sides of a border or between homeland and dependent territory, can both act as a barrier and an attraction to cross-border travel (Canally and Timothy, 2007; Spierings and van der Velde, 2013; Timothy, 2001c). For many people, the distinctive socio-cultural landscapes on the other side adds a level of charm and differentness that reinforces the idea of being somewhere else (Timothy and Canally, 2008). Despite the prevalent notion that different is good when seeking a travel destination, standard wisdom simultaneously suggests that transfrontier cultural differences should not be too stark, for a certain level of familiarity abroad is important because it helps travelers feel comfortable (Spierings and van der Velde, 2013). Thus, most tourists want different but not too different.

Exclaves are a nice laboratory for understanding this. Separateness and just a modicum of differentness have long been one of the draws of exclaves for people from the homeland. They can visit the exclaves, encircled by foreign soil, without having to experience 'foreignness'. They can 'go abroad' but remain in a piece of the homeland where they may share the language and food and where social attitudes are familiar. By the same token, visitors from the host state, or enclaving country, can cross an international border by visiting the enclavic territory and experiencing a limited degree of 'foreignness' without having to pass through tradititonal passport and customs inspections in either direction, which was a significant issue before the 1990s (Timothy, 1996). French customers then (and today) were keen shoppers in Spain, and in the Cerdanya Valley, a quick trip over Llívia's enclavic frontier gave passport-free and customs-free access to Spanish wines, spirits and Iberian ham. Similar situations pervaded among Swiss shoppers in Campione d'Italia and Dutch consumers in Baarle-Hertog.

British visitors to the Isle of Man, Jersey, Guernsey, or Gibraltar will be satisfied with the level of cultural difference that exists between the UK at the overseas territories or crown dependencies. They are each culturally, economically, and socially different enough to provide a 'foreign flair' but British enough that they also feel at home.

Distinct culture and language

Related to the point above, Baum (1997) emphasizes that linguistic and cultural differences wield considerable appeal in small, foreign territories. Timothy (1995, 2001c) notes how crossing borders can be part of the excitement of travel, in large part because of the cultural or linguistic differences that lie on the other side. Likewise, religious and culinary differences may add an additional layer of interest. Andorra proudly touts its natural Pyrenean cuisine in its print and online advertising, and its unique mountain gastronomy is considered one of its national points of pride, together with the fact that it is currently the only sovereign state where Catalan is the national language. Even San Marino's culinary heritage, which is essentially the same as Italian cuisine from neighboring Emilia-Romagna and Marche, includes a few exclusive specialty foods with a Sammarinese twist. Equally, the Sammarinese (Romagnol) dialect of Italian is the language of choice in the republic, which it proudly publicizes as part of its national identity. The Faroe Islands' unique language, Faroese, and inimitable sea-based gastronomy, which is uniquely different to that of mainland Denmark, are part of the total attraction base of the islands.

Likewise, the dying indigenous languages of Monaco (Monégasque), the Isle of Man (Manx), Jersey (Jèrriais) and Geurnsey (Guernésiais) are seeing a resurgence, partially through tourism. The presence of place names, road signs, and public notices printed in these endangered languages adds a deeper cultural appeal to these small and isolated places and helps contribute to their protection. The food of Llívia is decidedly Catalan as is the language and national identity. That it is entirely enclaved by France increases even further the enclave's solidarity with Catalonia on the Spanish side of the border.

Going 'Back in Time'

This subsection highlights what Baum (1997) called 'a slower pace of life' and 'back in time' characteristics of insular places and microstates. Some of the geographical entities discussed in this tome clearly have a sense of being back in time. Mount Athos, Greece, a theocratic autonomous region of Greece is a good example (della Dora, 2012). Its High to Late Middle Age monastic landscape, which only men can visit, is in many ways frozen in time. The Medieval architecture, the Early Middle Age rites and rituals that are still practiced are a UNESCO World Heritage Site and coveted religious destination (Alexopoulos, 2013). Likewise, the highly paternalistic order, which precludes females from visiting, hearkens back to medieval times when women were impeded from participating in administering the church and the state.

Another suitable example of this is the Sovietization of Transnistria and a few of the other shadow states in Eastern Europe. Because some of these unrecognized states attempted to secede just before the fall of the USSR and today function de facto independently of the states to which they legally belong (e.g. Georgia, Moldova and Azerbaijan), their last effective sovereign motherland was the USSR. This, together with a strong dependency relationship with Russia and a sustained 'frozen' sense of Soviet nostalgia, has helped maintain a Soviet-like cultural landscape in Transnistra that is one of the polity's main attractions. Statues of Lenin are still pervasive in parks and in the courtyards of government buildings (MacLean, 2013; Stan, 2018). Hammer and sickle insignias are everywhere, including on the national flag. Soviet-era busses and trams provide local transportation, most of the Socialist Classicist buildings remain from the Stalinist architectural period, and the shadow state continues to celebrate Soviet holidays (Şveţ, 2013). Many former Soviet officers have retired there, and the preferred language of the majority is Russian. Much of Transnistria's tourism product is geared toward maintaining and interpreting its museumified Sovietscapes. Neither of these examples, Mount Athos or Transnistria, would have a unique selling proposition as a tourist destination without their matchless geopolitical histories that take tourists 'back in time'.

Distinctive Niche attractions

Tourism marketers and scholars divide tourism into unique types or niches as a way of targeting specific market segments or as a means of being able to study behaviors, expectations and experiences. There are growing tourism niches that seek unusual places—destinations that are out of the way, marginal and in some cases of little interest to the mainstream tourist. These niche markets tend to be quite allocentric (Plog, 2001), meaning that they are adventurous and desire to travel off-the-beaten tourist path. Such people may see themselves as pioneers or conquerors in their efforts to collect places or count themselves as one of the few select who have visited (Amoamo, 2011). In this way, small places on the global margins may become mythical and legendary. Certain de facto states, microstates, autonomous regions and exclaves fit this bill.

The geographic characteristics of peripherality, insularity and compact size are conducive to developing distinct tourism types for which microstates and other territories become well known (Weaver, 2018). Although the chapter outlined some of the difficulties associated with smallness and marginality, such as lack of resources constricting a tiny polity's ability to develop multiple forms of tourism, smallness may also be seen as beneficial in this regard in compelling peripheral and small polities to develop one or two specific types of tourism as specialties for which they become well known. Day trips or overnight visa-free trips to Russia are the tourism product most associated with the quasi-autonomous Saimaa Canal. Andorra is famous throughout Europe for skiing and shopping. As mentioned above, Transnistria is known for its Soviet-era nostalgic landscapes and institutions. The insularity and political separateness of the Isle of Man and the Channel Islands have made these into popular tax havens and centers of banking. Because of their unique border situations and autonomy status, Samnaun, Livigno, Åland and Heligoland are above all else, some of Europe's most popular duty-free shopping magnets, reflecting the words of Harrison (2001, p. 4), 'Geographic isolation can be an economic advantage, enabling an island [or enclave] to cater for interests not easily satisfied on the adjacent mainland'.

Other Benefits of Small Geography

Besides the geographical characteristics proposed by Baum (1997) and others (e.g. Amoamo, 2011; Butler, 1993; Grydehøj, 2011; Timothy, 1996, 2001a, 2001b), there are other features of microstates, small territories, and marginal polities that are beneficial to tourism and development more broadly.

Social solidarity

Although demographic smallness can create problems, it also has benefits. Small populations in Europe are well educated overall, frequently having a higher literacy rate than their larger neighbors. Their populations are hardworking and industrious with a strong desire to see their countries succeed. These characteristics, together with lower taxes, have made microstates popular locations for international business headquarters. Although most of the microstates rely on day labor to fill tourism positions, many high-end management jobs are filled by Andorran, Monacan, Sammarinese and Liechtensteiner nationals. Given their tax haven status, educated population and willing-to-work workforce, with the exception of the Vatican, most of the microstates have more registered businesses than what their populations could support. This is evidence of a cohesive community that works together for the betterment of their small society.

The sense of solidarity is an important characteristic that seems to prevail in the European microstates, exclaves and territories, which is a strength for quality of life, human rights, and the development of tourism. Because of their long historical struggles to survive in the chasm between different cultures, in their fight to remain relevant, and their shared displacement, enclavers share a common mentality, 'a peculiar ingenuity... [shaped] by centuries of having to deal with not one, but two state bureaucracies at a time' (Vitaliev, 2008). As borderlanders, exclave residents have become accustomed to living an interstitial 'in-betweenness' where they have had to resist pressure to assimilate, as well as negotiate their personal, community and national identities (Berger, 2020; Gelbman and Timothy, 2011). In this process, through adversity they have strengthened their solidarity on the margins of the state.

Such common struggles are pervasive in de facto states, dependencies, territories and microstates. Exclaves' interstitial living spaces and borderland conditions frequently engender an identity crisis and sense of unease that simultaneously connects them and isolates them (Bjelland, 2016). However, this geography-induced predicament generates a greater sense of geographical and social solidarity. 'People associating themselves with the local territory rather than with the country as a whole seems to be strong in an enclave in comparison with the mainland and the surrounding state. This local allegiance often outweighs the national or any other component of multiple identities' (Vinokurov, 2007, p. 117). Jansson (2012) demonstrates Åland islanders' in-betweenness. They are neither Finnish nor Swedish, but they are more Swedish than Finnish but still they are Ålanders first. One Campionesi told Vitaliev (2008, p. 112) 'I don't know who I am...in Switzerland, they take me for an Italian, whereas in Italy they dismiss me as a Swiss'. Likewise, although Büsingers are Germans, they are socially and psychologically more oriented towards Switzerland (Musiał-Karg, 2012) and have in the past desired (96% in the early 1900s) to be assimilated into Switzerland. Büsingen's German citizens celebrate Swiss holidays, preferred over German holidays because, according to one local official, 'Our spirit and heart are Swiss' (quoted in Bleiberg, 2019, n.p.). Many Jungholzers feel more German than Austrian (Vitaliev, 2008), whereas Llívians are unequivocally Catalan, but not Spanish. In Baarle, local identity as 'Baarlenaars' is in many ways more important than national identity as Dutch or Belgian (Vinokurov, 2007, p. 136). These examples align closely with the assertion of Vinokurov (2007, p. 117): 'Sentiments of political and cultural affiliation do not necessarily coincide'. In many exclaves the border complexity and the struggles and advantages it has caused over generations, have strengthened local identity, interdependence, and local-scale solidarity.

Observers have suggested that small populations usually translate into greater social homogeneity and solidarity (Armstrong and Read, 1994, 1998; Bertram, 2007; Conway, 1998). In the case of Europe's microstates, this is true. San Marino and Liechtenstein, for example, are true nation-states. This tends to unify their populations in many cultural and political ways. Small populations that have had to rely heavily on one another in the face of the difficulties of political marginality tend to have a stronger sense of national identity. This is very evident in the microstates, the de facto states, the Channel Islands, the Isle of Man, Gibraltar and Åland. This level of homogeneity is easier to work with in developing tourism plans, tourism-related legislation, and marketing campaigns.

This leads to easier governance, higher levels of participation and more localized control. Many commentators believe this to be a political advantage because more citizens are better acquainted with politicians, they understand the issues more thoroughly, and because of their familiarity with governors and local problems, they can more democratically influence legislation and decision making though majority rule or consensus (Armstrong and Read, 1998; Anckar, 2008; Bertram, 2007; Niikawa and Corcaci, 2016; Veenendaal, 2013a, b, 2020). In microstate politics, interpersonal relations and informal dynamics between politicians and the electorate are run-of-the-mill occurrences (Baldacchino, 2012; Veenendaal, 2015a, 2016b). Despite its positive potential in tourism decision making, Veenendaal (2015b) warns that this close-knit system of democracy has potential to result in conflicts of interest and nepotism.

Flexibility and stability

It is a common belief that microstate populations, policies and laws are more flexible and stable than in larger states where more coordination is required (Thürer, 1998) and where there is more dissent. From Veenendaal's (2015b) perspective, the national governance of microstates is comparable to local-level governance in large countries, which means that governance in general is by nature a grassroots endeavor. In the domain of sustainable tourism planning, this is a strength. Through grassroots, participatory development, projects may be less encumbered by diverging interests, increasing the flexibility and stability of a given development program. As noted in the previous section, when a small population stands together in solidarity, including in tourism development, flexibility is better insured and protected. Policies and development frameworks that exhibit more flexibility are more adaptable to changes that might occur in the planning process or when a plan is being implemented. In addition, microstates are adept at cross-border cooperation in foreign policy (Neumann and Gstöhl, 2006). This skill may be utilized to minimize development costs by initiating shared actions, sponsorships or joint infrastructure development as has been done in many other interstate tourism development efforts throughout the world.

Europe's microstates are generally more stable than the countries that surround them—a stability largely maintained because of their higher level of national solidarity and homogeneity. This is one of the reasons they have been so successful in economic development and why their tourism thrives. Tourism in small, relatively unknown destinations, such as the microstates, requires political stability, high standards of living, strong economic growth, a good infrastructure and a unified population (Vítová *et al.*, 2019).

Conclusion

The overarching issue in relation to tourism and microstate studies is power relations. Powerlessness overall or in certain areas leads to undeniable vulnerabilities that derive naturally from small size, small populations and marginal locations, yet these very same characteristics that would seem to disempower small states and territories in strictly geopolitical, economic or social terms, simultaneously empower them in the tourism domain! Tourism has a way of defying the normative models of peripheral (under) development and breaking the bands of the status quo. Although miniscule territory and small population, as well as marginal location and inaccess, can raise overwhelming barriers to sustainable and meaningful development, they are a large part of what makes microstates and other European territories attractive for tourism in the first place.

Many vulnerabilities derive from smallness and marginality, not least of which is overdependence on tourism and on the exogenous other. This may entail putting too many eggs in the tourism basket, making small and peripheral territories increasingly susceptible to the delicate vagaries of global economics, political turmoil and even health pandemics. It also renders some of them defenseless against industry and market leaders abroad whose unilateral decisions affect their own domestic supply and demand. In the words of Bailey (1998, p. 310), 'Marginality breeds dependence'. Dependence on exogenous forces leads to a reduced ability to determine their own futures. Wherever a power gap exists, another entity will move in to fill it—a larger neighbor, a patron state with ulterior motives, a host country that may have different goals and priorities, or a colonial metropole with dependency interests.

Just like people, microstates, enclaves and de facto states wish to be acknowledged and accepted by their peers. In the world of diplomacy and geopolitics, political bullying and neglect occur quite regularly, and it is the smaller states and regions that are most often on the receiving end. De facto states are in a perpetual state of non-recognition, and microstates are sometimes categorically dismissed as incapable imposters. Exclaves suffer a similar fate but at the domestic level where their homelands sometimes forget they exist, or intentionally ignore them to avoid complicated situations. Many enclavers have voiced a sense of home state abandonment, which frequently sows discontent and a desire to unify with the host.

Despite the vulnerabilities brought about by smallness and liminality, small and unusual areas are thriving tourist destinations. Their spatial, political and/or social isolation determines much of their appeal, successfully accentuating their differentness in culture, heritage and history, and forming unique tourism products that would not likely emerge in larger states. One of the deepest characteristics of successful tourism is the idea of differentness. People leave home to experience different places, places beyond their ordinary. Europe's anomalous regions are atypical and extraordinary, which is the foundation of their tourism success.

12

Perspectives on Tourism and Small Polities in Europe

This book is foremost about tourism's relationships with the political geography of smallness, marginality, separation, overdependence, political vulnerabilities, and pan-Europeanism. Tourism is one of the region's most significant economic sectors, and Europe is the world's most visited region. Yet, not every place in Europe has equal opportunities to benefit from tourism. In fact, many fascinating places do not even appear on the normative tourism radar, because they are small, unknown or politically and socially powerless.

This book builds upon the growing literature on the political geographical elements of place and how tourism is modified by, or modifies, certain places positively or negatively. The basic unit of analysis is the sovereign state, but each state's territory looks different and sometimes behaves differently in unique contexts throughout the world.

Notwithstanding their legal status as sovereign, independent countries, the microstates examined here—Andorra, Liechtenstein, Monaco, San Marino and Vatican City—all are in dependency relationships with a larger caregiver state. Their history of ineffective governance due to their small area and population necessitated their intertwining associations with a patron state to represent their interests on the global stage. In the 1990s, however, four of the five microstates were endorsed to enter the society of states, in this case the United Nations, and since that time, some of them have become emboldened to begin standing on their own at the supranational level. Because of their small physical size, which is not conducive to satisfying all the requirements of EU membership (e.g. freedom of movement and abode), four of the five states have joined the European common market through other legal mechanisms that do not require full membership in the European Union. Their access to the single market has important repercussions for tourism, as outlined in Chapter 2.

Each one of Europe's mini microstates has its own personality and tourist appeal, but what they all have in common is as follows. First, their sovereignty empowers them to govern their own tourism industries with minimal external interference. Second, their separateness in political, social and economic terms from their protector states enables them to develop unique sorts of tourism (e.g. business travel, gambling and shopping) that draw visitors from nearby and afar and develop a quaint and appealing personality. Third, their borders are defined, delimited and demarcated, thereby ensuring a uniquely developed national identity that harbors an inimitable sense of patriotism and solidary not found in many of the continent's larger countries. Finally, they all have certain political, economic and social vulnerabilities that require them to associate with a larger protector state.

Autonomous regions have more independence than any other parts of the state to which they belong. Many have selected tourism to represent that autonomy and cash in on the rewards. The British crown dependencies are independent within themselves and must give permission for the UK to legislate on their behalf. Some autonomous areas are part of the European Union but have received special dispensation to reduce their tax rates (e.g. Åland and Heligoland) and establish some of their own regulations that would not normally be permitted within the EU (e.g. Mount Athos permitting only men to enter and requiring a special permit even though it is also under the auspices of the Schengen Agreement) as a means of supporting a tourism-heavy economy. Other autonomous areas are not in the EU or Schengen (e.g. the Faroe Islands and Svalbard) and are therefore freer to direct their own internal tourism-related matters.

Exclaves of several varieties dot the geopolitical landscape of Europe, creating advantages and disadvantages for both home and host states. These have a much lower level of self-determination, as they are integral parts of home states without autonomous status. However, their separation and distance from the motherland has, in the past, created practical autonomous-like conditions where they have been left to their own devices with regard to tourism and other sectors. This has allowed the industry to thrive in many of Europe's tiniest territories, several of which have become prominent regional tourist attractions and destinations.

Aside from embassies and consulates, extraterritorial places are few. In these special conditions, the occupied space remains the sovereign territory of the host state (not the occupier), but local laws either do not apply or are limited. This can create unique opportunities for tourism and jurisdictional conflicts if not managed properly. As an example of opportunities, sailing on the Finnish-leased and controlled Saimaa Canal in Finland and Russia is one of only a few ways in which most tourists can visit Russia without a visa. Although the eastern half of the canal is entirely in Russia, many of its laws are suspended in the canal's leased area according to international treaty.

Unrecognized pseudo-states and micronations are also tourist destinations. All of Europe's pseudo-states (de facto states) are associated with conflict. Although this raises barriers to mass tourism, there is a type of adventurous tourist who seeks opportunities to visit places of stark otherness and places of conflict. These curiosity seekers form the biggest markets for the de facto states, although in Cyprus, visiting the occupied north is no longer a dangerous sense of adventure but rather a mainstream tourist activity. Micronations are the least legitimate micro-territories in the world, as they are not recognized and in most cases have few legal grounds upon which they can base their claims to sovereignty or independence. Yet, their novelty value is undeniable and attracts people who either collect unusual places or who happen upon them while holidaying nearby.

Geopolitics affects everyone, from the food we eat to the travel choices we make, yet few scholars have dug into the bones of political geography to provide an alternative political understanding of tourism. Those who have done this have dealt with mainstream countries in Western and Eastern Europe, although Hall's (2017b) recent collection of case studies provides useful insight into the geopolitics of tourism in some unusual proto-states, conflict areas and coastal outliers in Eastern Europe, such as Crimea, Kosovo and Kaliningrad. Hall's efforts are worth trailing and building upon as we try to comprehend the multitudes of state types, evolutionary processes, and the jurisdictional reach of sovereignty. This book follows the work of Hall (2004, 2017b), which spotlights transitional geopolitics in Eastern Europe in relation to tourism, and Timothy's (2001c) work on political borders and tourism.

Despite these efforts, heretofore, nobody has systematically described the smallest and most unusual polities in Europe within the framework of tourism and political geography. Because of their locations, regional prominence, and volume, they deserve to be considered as special territories in a dynamic world region where traditional notions of independence and sovereignty are being challenged today more than anywhere else on earth. Nonetheless, doing research in such areas is not easy.

One of the premier challenges of doing research in microstates, enclaves, autonomous regions, territories, dependencies, de facto states, extraterritorial areas, and micronations is not so much about political vulnerabilities or security threats, but rather a lack of data and research opportunities. When microstates and similar entities complain of non-recognition in the world of supranationalism and diplomacy, the same can be said of tourism representation. Official tourism statistics compiled by the World Tourism Organization or the World Travel and Tourism Council typically focus only on sovereign states and a limited number of territories (e.g. Hong Kong, New Caledonia and Martinique). However, although formal states and territories constitute a large portion of the earth's area, they do not encompass all of it. Likewise, while officially recognized states and territories represent the majority of global tourism, they certainly do not represent it all. Thus, millions of visits and millions of euros in expenditures go unrecorded each year, because data is reported at the state or territorial level.

Even efforts to understand certain patterns, such as microstate outbound travel, are grossly inaccurate because of the lack of border controls and ignorance or inefficiencies on the part of receiving countries who are unable to differentiate between Sammarinese and Italians, or Monacans and French. Funded research opportunities are difficult in unrecognized states, as sanctions and soured political relationships frequently build powerful administrative obstacles to funding and collaboration. Likewise, many small and anomalous entities lack funding to provide research opportunities beyond industry-specific marketing data.

Despite these challenges, these unique political geographic settings provide bounteous opportunities for additional research, whether funded or not. Multitudes of questions can be raised about the internal dynamics of sovereignty in microstates and how these are challenged or empowered in the complexity of tourism. Likewise, comparative studies between similar types of states or proto-states would be useful in understanding relationships between sovereignty and tourism growth inside different political regimes.

Although general patterns of patron-to-client travel and vice versa have been described, developing an understanding of other market segments, particularly those who are not part of the traditions of mass tourism, could be fruitfully brought to bear on knowing more about the ways in which geopolitical anomalies attract different tourist markets.

As comprehensive as this book tried to be, not all political geographical anomalies in Europe outside the normative state were included. There are other exceptional territories worth mentioning and which would benefit from future empirical research. This book addressed three areas—the Turkish Republic of Northern Cyprus, Luhansk and Donetsk, which satisfy the definition of occupied territories, but they are also de facto states. In 2014, as part of its eastern Ukraine offensive, Russia occupied and unilaterally annexed the Crimean Peninsula. This move has been condemned internationally and goes beyond mere military occupation. To occupy and annex part of a sovereign state for irredentist purposes is generally not how modern states gain international sympathy. Illegally annexed territories are few in Europe, but Crimea is a good example of one that has major associations with tourism. Even before 2014, Crimea was a popular destination, but its annexation by Russia has elevated its stature within that country into an even more desirable destination (Ivanov *et al.*, 2016). Tourism there is thriving now that it is considered a domestic destination once again, and Russia is planning to undertake major bridge-building projects to link the peninsula physically with the Russian mainland. In modern times in Europe, there are few such annexed places, which differ from de facto, breakaway states in several regards. The most notable differences is that they do not claim to be independent countries on their own, and they lack the external patronage that fuels the continuation of the unrecognized quasi-states.

A second territorial peculiarity in Europe that could merit additional research attention is the condominium. In international law, condominiums are places over which sovereignty is shared by more than one sovereign state. These are very rare today but were more common in the distant past. It is rather surprising that such a place might exist, given the jealousness with which territorial sovereignty is so brazenly defended. Two current condominiums in Europe have tourism potential, although only one is currently being utilized for tourism. The first is Pheasant Island, a small island in the Bidasoa River on the border of France and Spain. The Treaty of the Pyrenees, which established the current border of France and Spain, was signed on Pheasant Island in 1659. The very same treaty stipulated that the island's sovereignty would be shared between the two countries in perpetuity. Each country alternates control of the island every six months for the purpose of cleaning and maintenance, including maintaining the historic monument that commemorates the treaty (Sánchez Arsuaga, 2018). The island and its historical marker has some tourism potential, especially among Spanish and French tourists, for the events on the island 361 years ago helped confirm their identities as French and Spanish. However, the island is treated as a military area and is strictly off limits to everyone except municipal maintenance workers.

The second condominium comprises the Moselle, Sauer and Our Rivers where they form different segments of the boundary between Luxembourg and Germany. The sovereignty of Germany and Luxembourg ends at their own riverbanks, leaving the river itself under the joint sovereignty of the two neighbors (Khan, 2004). A person on a bridge or boating on any of these border rivers is physically and legally in two countries at once. Pheasant Island and the Luxembourg-Germany condominium (*deutsch-luxemburgische Hoheitsgebiet*) are the only places in Europe where this is possible. Although most recreational users of the river would not be aware of this unique legal status, the condition certainly has implications for using the rivers for tourism and recreation. Tourism planning with regard to the waterway, water usage, and recreation facilities must be approved by both countries and would be reviewed using a separate set of laws and ordinances.

This book also touched upon secessionism, or efforts by a people or region to separate from the state to which it belongs in order to form its own independent country. This is the crusade that brought many of Europe's de facto states into being when the USSR dissolved. There are dozens of secessionist movements worldwide, but some of the strongest are in Europe. Kosovo's February 17, 2008, declaration of independence from Serbia remains a significant point of contention between Serbia, Kosovo and the 97 (half of the UN membership) states that have recognized Kosovo's independence. Scotland's ongoing efforts to secede from the UK appear to be gaining strength. Catalonia's formidable independence movement has been a significant concern for Spain, which relies on the autonomous region for its strong economic contribution to state coffers. The Basque separatist movement has calmed down in recent years after Spain granted the region more autonomy, but other parts of Spain, France, the UK, Italy and Denmark are contemplating the advantages and weighing the disadvantages of going it alone in the global network of sovereign states.

Separatism can have deeply significant implications for tourism and other mobilities. In the process of secession, domestic boundaries become international borders overnight. Based on lessons from the acquisition of independence by Namibia (1990), the Czech Republic and Slovakia (1993), East Timor (2002), Montenegro (2006) and South Sudan (2011), acquiring sovereignity is costly and complicated. Customs and passport inspections are erected immediately at state borders. Some roads are truncated and cut off, even for local traffic. New state services must be created: law enforcement, postal services, currency regulations (and sometimes a new currency), national park authorities, ministries of tourism, public works, military authorities, and transport authorities. New diplomatic relations must be negotiated and embassies build abroad. New passports have to be printed and distributed, and the new state must begin to build an independent tourism brand.

Similarly, but on a supranational level, the UK's January 31, 2020, divorce from the European Union will likely bring unforeseen consequences, which has stoked fears in many of the UK's constituent countries, overseas territories, and dependencies. There are rumblings of other EU exits among certain member countries, but some of them plan to wait on their withdrawal until they have a better grasp of its implications by watching what happens in Great Britain.

Not all challenges to small states and territories are purely geopolitical, but a mix of geopolitics and environmental conditions. Climate change is affecting the entire world, with rising temperatures, increasing numbers and intensities of weather events, higher wildfire risks, rapid ice melt in the arctic and Antarctic regions, and reduced levels of precipitation. Small island microstates in the Pacific and Indian Ocean are especially in danger from rising seas (Belle and Bramwell, 2005). In Europe, rising temperatures also affect tourism in several ways. Icemelt and reduced snowfall in Svalbard is having a significant impact on polar bear populations and other resources that attract tourists to the region. It is also diminishing the polar bears' food sources (i.e. seals and fish), forcing the animals to inch closer to human civilizations to find food. As well, a lower volume of snowfall is a major concern for the ski industries of Andorra, Liechtenstein, Kleinwalsertal, Jungholz, Livigno and Samnaun, and these small states and pene-exclaves are among the hardest hit snow destinations in Europe (Pons-Pons *et al.*, 2012; Steiger *et al.*, 2019).

Publications about sustainable tourism have amassed in the thousands since the 1980s with sweeping recommendations for greener, more responsible, more equitable, and more participatory approaches to tourism development. Yet, as noted earlier in another regard, these clarion calls for better tourism are based on studies in normative places with conventional governance frameworks. Perhaps we need to step back and look at atypical polities under alternative governance, because as Wilkinson (1994) declared over a quarter century ago, sustainability sometimes looks different in small states than it does in large states. Likewise, what does it look like in Europe vs Asia or the Pacific? This is an important question to raise, for as Wivel (2020, p. 110) notes, 'The small states in Europe are privileged. They are typically stronger and more resilient than small states in other parts of the world'. What does sustainable tourism or even sustainable development look like in quasi-states that are struggling to remain on the map? How does sustainability play out in microscopic polities that are isolated from their homelands and may sometimes face obstruction by the state that surrounds them? What if sustainable tourism in a microstate is dynamically or conceptually different than sustainable tourism in the patron state that oversees it? These and many other such questions remain unanswered, but they are important to consider.

At the time of writing, the COVID-19 pandemic is still in full force. In an effort to contain the spread of the Coronavirus, many European countries have closed their borders almost entirely, only allowing essential travel across state frontiers. This demonstrates that although member states have turned over much of their autonomy, and even sovereignty, to the European Commission, they still have enough to exercise their rights to close borders as needed to protect their citizens. During the pandemic, Lithuania, Latvia and Estonia together decided to create a 'travel bubble' early on to enable free movement across their common borders. Norway and Denmark allowed cross-border travel in June 2020, as did Switzerland, Germany, France and Austria (BBC News, 2020). The rise of this dreaded disease has shown the power of the state to halt tourism and then to enable its revival. For some microstates, enclaves, islands, and other possessions, this exercise in border control has created a double isolation effect, which may be even harder to recover from than political disruptions.

In terms of COVID-19, according to Dellanna and Monella (2020), San Marino was the worst affected country on earth in terms of deaths per capita at the end of April 2020. To slow the spread of Coronavirus, the microstate closed its border with Italy at almost every crossing point and rural road, leaving only a few open for essential travel under strict surveillance. Andorra's borders with France and Spain were sealed but began opening again at the end of May 2020. This self-imposed double isolation is particularly acute in these tiny landlocked countries, as all imports, exports and travel must take place through Italy, France or Spain, and a microstate's solutions to a pandemic may need to be found within the confines of its tiny frontier. Liechtenstein's border with Austria closed for a time, but its border with Switzerland remained open during the crisis. So, even between Europe's microstates, there are different cross-border dynamics at play that either reinforce a mini state's autonomy or its over-reliance on a larger master.

This book is foremost about power relations between small and large, near and far, dependent upon and depended on, in several atypical territories of Europe. Tourism both benefits and suffers simultaneously from these power relations, yet tourism, if properly planned and managed, can be an effective tool in helping empower the disempowered and enlarge even the smallest of places. It has been suggested that tourism, perhaps more than any other industrial sector, can help 'overcome the tyranny of size' (Vanegas and Croes, 2003, p. 315).

Final Word

During the past 25 years, I have visited almost every one of the unique places described in this book both for research and for the pure sake of curiosity. Hundreds of residents, local administrators, tourism sector managers and planners, border officials, and tourists have been interviewed, surveyed with questionnaires, and observed over the years to gauge their behaviors, interests and motives. In addition, I have read thousands of academic publications, popular media stories, travel reports, promotional ads, and websites. Through this work of the past quarter century, it has become abundantly clear that the human geopolitical environment is far more complicated than most of us realize. When it comes to the geographical, political and legal organization of the world, there are no norms, only generalizations and exceptions. Every place is unique, and each locality has its own challenges and opportunities.

Since the mid-1900s, tourism has surfaced and magnified these challenges and enriched the opportunities. Tourism is a double-edged sword, but almost every place on earth desires to have it. The small and unusual geopolitical entities described in this book are no exception. Some of their political geographical conditions are favorable for developing a strong tourism sector. Others present insurmountable obstacles. What we do know, is that we know relatively little about the dynamics of political geography and how it intersects with tourism. Much more work is needed on these very issues, as well as peace, discontent, separatism, irredentism and other political geographical concepts that are important for tourism and community well-being. Small states, tiny territories, and other anomalous entities are suitable laboratories for understanding these dynamics, because 'due to their physical vulnerability, small states have always been interested in a 'civilization' of international relations' (Goetschel, 1998, p. 20). This book demonstrates clearly that everyday life in microstates, exclaves, and similar polities is an enduring exercise in international relations!

References

Åkermark, A.S., Heinikoski, S. and Kleemola-Juntunen, P. (2018) *Demilitarisation and International Law in Context: The Åland Islands*. Routledge, London.

Åland Post (2020) The special status of the Åland Islands. Available at: https://www.alandpost.com/axla-logistics-e-commerce-fulfilment/special-status-aland-islands (accessed 19 February 2020).

Ålands Lagting (2020) The autonomous Parliament of Åland. Available at: https://www.lagtinget.ax/ (accessed 22 March 2020).

Alexopoulos, G. (2013) Management of living religious heritage: who sets the agenda? The case of the monastic community of Mount Athos. *Conservation and Management of Archaeological Sites* 15(1), 59–75.

Alexopoulos, G. and Fouseki, K. (2016) Gender exclusion and local values versus universal cultural heritage significance: the Avaton debate on the monastic community of Mount Athos. *European Journal of Post-Classical Archaeologies* 6, 235–254.

Algieri, B. (2006) International tourism specialisation of small countries. *International Journal of Tourism Research* 8(1), 1–12.

Alipour, H., Vaziri, R.K. and Ligay, E. (2011) Governance as catalyst to sustainable tourism development: evidence from North Cyprus. *Journal of Sustainable Development* 4(5), 32–49. DOI: 10.5539/jsd.v4n5p32.

Allegrezza, S., Briguglio, L. and Spolaore, E. (2019) Competitiveness of small states: guest editorial. *Small States & Territories* 2(1), 3–8.

Altinay, L. and Bowen, D. (2006) Politics and tourism interface: the case of Cyprus. *Annals of Tourism Research* 33(4), 939–956.

Amoamo, M. (2011) Remoteness and myth making: tourism development on Pitcairn Island. *Tourism Planning & Development* 8(1), 1–19.

Amoamo, M. (2018) More thoughts on core-periphery and tourism: Brexit and the UK overseas territories. *Tourism Recreation Research* 43(3), 289–304.

Anckar, D. (2008) Microstate democracy: majority or consensus; diffusion or problem-solving? *Democratization* 15(1), 67–85.

Andorra Turisme (2020) What to do in Andorra? Available at: https://visitandorra.com/en/# (accessed 3 January 2020).

Angus, C., Holmes, J. and Meier, P.S. (2019) Comparing alcohol taxation throughout the European Union. *Addiction* 114, 1489–1494.

Ankre, R. and Nilsson, P-Å. (2016) Remote yet close: the question of accessibility in the Faroe Islands. In: Baldacchino, G. (ed.) *Archipelago Tourism: Policies and Practices*. Ashgate, Farnham, pp. 137–145.

Anthony, E.J. (1994) Natural and artificial shores of the French Riviera: an analysis of their interrelationship. *Journal of Coastal Research* 10(1), 48–58.

Armstrong, H.W. and Read, R. (1994) Micro-states, autonomous regions and the European Union. *European Urban and Regional Studies* 1(1), 71–78.

Armstrong, H.W. and Read, R. (1995) Western European micro-states and EU autonomous regions: the advantages of size and sovereignty. *World Development* 23(7), 1229–1245.

Armstrong, H.W. and Read, R. (1998) Trade and growth in small states: the impact of global trade liberalisation. *The World Economy* 21(4), 563–585.

Armstrong, H.W. and Read, R. (2000) Comparing the economic performance of dependent territories and sovereign microstates. *Economic Development and Cultural Change* 48(2), 285–306.

Armstrong, H.W. and Read, R. (2003) Small states, islands and small states that are also islands. *Studies in Regional Science* 33(1), 237–260.

Armstrong, H., De Kervenoael, R.J., Li, X. and Read, R. (1998) A comparison of the economic performance of different micro-states, and between micro-states and larger countries. *World Development* 26(4), 639–656.

Azienda di Promozione e Sviluppo Turistico (2020) Livigno: feel the Alps. Available at: https://www.livigno.eu/en/ (accessed 30 March 2020).

Azienda Turistica (2020) Welcome to Campione d'Italia: let yourself be enchanted by the beauty of our exclave. Available at: http://www.campioneitalia.com/wordpress/ (accessed 26 April 2020).

Bahcheli, T. and Noel, S. (2010) The political evolution of Northern Cyprus and its effect on Turkish-Cypriot relations with Turkey. *The Cyprus Review* 22(2), 139–148.

Bailey, P. (1998) Faroe, Orkney, Gran Canaria: case studies in the geography of marginal Europe. *Geography: Journal of the Geographical Association* 83(4), 309–321.

Baldacchino, G. (1993) Bursting the bubble: the pseudo-development strategies of microstates. *Development and Change* 24(1), 29–52.

Baldacchino, G. (1994) Peculiar human resource management practices? A case study of a microstate hotel. *Tourism Management* 15(1), 46–52.

Baldacchino, G. (1997) A clash of human resource management cultures: a micro-state case study. *International Journal of Human Resource Management* 8(4), 506–518.

Baldacchino, G. (2012) Islands and despots. *Commonwealth and Comparative Politics* 50(1), 103–120.

Baldacchino, G. (2015) *Entrepreneurship in Small Island States and Territories*. Routledge, London.

Baldacchino, G. (2018) Mainstreaming the study of small states and territories. *Small States & Territories* 1(1), 3–16.

Baldacchino, G. (2019) The competitiveness of small states: insights on flexible specialization. *Small States & Territories* 2(1), 41–54.

Baldacchino, G. and Pleijel, C. (2010) European islands, development and the cohesion policy: a case study of Kökar, Åland Islands. *Island Studies Journal* 5(1), 89–110.

Baldacchino, G. and Wivel, A. (2020) Small states: concepts and theories. In: Baldacchino, G. and Wivel, A. (eds) *Handbook on the Politics of Small States*. Edward Elgar, Cheltenham, pp. 2–19.

Bartmann, B. (2002) Meeting the needs of microstate security. *The Round Table* 91(365), 361–374.

Bartmann, B. (2012) From the wings to the footlights: the international relations of Europe's smallest states. *Commonwealth & Comparative Politics* 50(4), 535–554.

Baum, T. (1996) Tourism and the Åland Islands. *Progress in Tourism and Hospitality Research* 2(2), 111–118.

Baum, T. (1997) The fascination of islands: a tourist perspective. In: Lockhart, D.G. and Drakakis-Smith, D. (eds) *Island Tourism: Trends and Prospects*. Pinter, London, pp. 21–35.

BBC News (2020) Coronavirus: Denmark and Norway exclude Sweden from tourism. *BBC News*, 29 May 2020. Available at: https://www.bbc.com/news/world-europe-52853556 (accessed 29 May 2020).

Beattie, D. (2012) *Liechtenstein: A Modern History*. Frank P. van Eck Publishers, Triesen, Liechtenstein.

Bechtel, D. (2004) A holiday in Samnaun is anything but taxing. *SwissInfo.ch*, January 21. Available at: https://www.swissinfo.ch/eng/a-holiday-in-samnaun-is-anything-but-taxing/3729256 (accessed 15 January 2020).

Belle, N. and Bramwell, B. (2005) Climate change and small island tourism: policy maker and industry perspectives in Barbados. *Journal of Travel Research* 44(1), 32–41.

Berger, S. (2010) The study of enclaves–some introductory remarks. *Geopolitics* 15(2), 312–328.

Bernouin, M. (2020) Crowded Monaco reclaims land to build more luxury flats with sea view. *Reuters*, 15 January. Available at: https://www.reuters.com/article/us-monaco-extension/crowded-monaco-reclaims-land-to-build-more-luxury-flats-with-sea-view-idUSKBN1ZE1Y4 (accessed 30 March 2020).

Bertram, G. (2007) *The Welfare State in Microstates and Small States*. United Nations Research Institute for Social Development, Geneva.

Beveridge, E. and O'Gorman, K.D. (2013) The Crusades, the Knights Templar and Hospitaller: a combination of religion, war, pilgrimage and tourism enablers. In: Butler, R. and Suntikul, W. (eds) *Tourism and War*. Routledge, London, pp. 39–48.

Bézias, J.R. (2007) Les Alpes-Maritimes et la crise franco-monégasque de 1962. *Cahiers de la Méditerranée* 74, 321–336.

Bickerton, C.J. (2012) *European Integration: From Nation States to Member States*. Oxford University Press, Oxford.

Birch, J.D. (1998) *Pilgrimage to Rome in the Middle Ages: Continuity and Change*. Boydell Press, Woodbridge.

Bjelland, M. (2016) Borderland lifeways in the American exclave of Point Roberts, Washington. *Geographical Review* 106(4), 516–538.

Bleiberg, L. (2019) Germany's tiny geographic oddity. BBC Travel. Available at: http://www.bbc.com/travel/story/20190922-germanys-tiny-geographic-oddity (accessed 15 October 2019).

Bloom, J. (2017) Free trade area, single market, customs union—what's the difference? BBC News, 14 August 2017. Available at: https://www.bbc.com/news/business-36083664 (accessed 5 December 2019).

Bolli, M. (1954) Die Enklave Büsingen. *Geographica Helvetica* 9(1), 225–313.

Borthakur, A. (2017) An analysis of the conflict in the Ferghana Valley. *Asian Affairs* 48(2), 334–350.

Brau, R., Lanza, A. and Pigliaru, F. (2007) How fast are small tourism countries growing? Evidence from the data for 1980–2003. *Tourism Economics* 13(4), 603–613.

Briguglio, L. (1995) Small island developing states and their economic vulnerabilities. *World Development* 23(9), 1615–1632.

Briguglio, L. (2016a) Introduction: small states and the European Union. In: Briguglio, L. (ed.) *Small States and the European Union: Economic Perspectives*. Routledge, London, pp. 1–27.

Briguglio, L. (2016b) The Maltese economy: structure, performance and European Union membership. In: Briguglio, L. (ed.) *Small States and the European Union: Economic Perspectives*. Routledge, London, pp. 184–201.

Briguglio, L. and Vella, M. (2019) The small states of the European Union and the resilience/competitiveness nexus. *Small States & Territories* 2(1), 55–68.

Britton, S.G. (1982) The political economy of tourism in the Third World. *Annals of Tourism Research* 9(3), 331–358.

Britton, S.G. (1987) Tourism in small developing countries. In: Britton, S. and Clarke, W.C. (eds) *Ambiguous Alternative: Tourism in Small Developing Countries*. University of South Pacific, Suva, pp. 167–194.

Burnett, G.W. and Uysal, M. (1991) Dominica—geographic isolation and tourism prospects. *Tourism Management* 12(2), 141–145.

Büsser, R. (2016) Parlamentarisches Policy-Agenda-Setting des monarchischen Staatsoberhaupts in Liechtenstein: Drei Fallbeispiele. In: Wolf, S. (ed.) *State Size Matters: Politik und Recht im Kontext von Kleinstaatlichkeit und Monarchie*. Springer, Wiesbaden, pp. 59–82.

Butler, G. and Barrett, G. (2018) Europe's 'Other' open-border zone: The Common Travel Area under the shadow of Brexit. *Cambridge Yearbook of European Legal Studies* 20, 252–286.

Butler, P. and Morris, C. (eds) (2017) *Small States in a Legal World*. Springer, Cham, Switzerland.

Butler, R.W. (1993) Tourism development in small islands: past influences and future directions. In: Lockhart, D.G., Drakakis-Smith, D. and Schembri, J. (eds) *The Development Process in Small Island States*. Routledge, London, pp. 71–91.

Butler, R.W. and Mao, B. (1996) Conceptual and theoretical implications of tourism between partitioned states. *Asia Pacific Journal of Tourism Research* 1(1), 25–34.

Calvini, N. (1992) *Il Principato di Seborga: Un millennio di storia*. Dominici, Rome.

Canally, C. and Timothy, D.J. (2007) Perceived constraints to travel across the US-Mexico border among American university students. *International Journal of Tourism Research* 9(6), 423–437.

Canavan, B. (2012) The extent and role of domestic tourism in a small island: the case of the Isle of Man. *Journal of Travel Research* 52(3), 340–352.

Canavan, B. (2013) Send more tourists! Stakeholder perceptions of a tourism industry in late stage decline: the case of the Isle of Man. *International Journal of Tourism Research* 15(2), 105–121.

Canavan, B. (2014) Sustainable tourism: development, decline and de-growth. Management issues from the Isle of Man. *Journal of Sustainable Tourism* 22(1), 127–147.

Canavan, B. (2015a) Managing tourism decline: insights from the Isle of Man. *Managing Sport and Leisure* 20(3), 174–190.

Canavan, B. (2015b) Marketing a tourism industry in late stage decline: the case of the Isle of Man. *Cogent Business & Management* 2(1), 1004227.

Canavan, B. (2017) Tourism stakeholder exclusion and conflict in a small island. *Leisure Studies* 36(3), 409–422.

Cappucci, M. and Zarrilli, L. (2008) Cultural landscape and national iconography: nation and identity in Armenian tourism. *GeoJournal of Tourism and Geosites* 1(1), 38–47.

Cardinali, M. (1982) *San Marino e la suo storia*. Kina Italia, Milan.

Carlsen, J. (2017) A systemic approach to tourism in island states. In: Ghosh, R.N., Siddique, M.A.B. and Gabbay, R. (eds) *Tourism and Economic Development: Case Studies from the Indian Ocean Region*. Routledge, London, pp. 94–103.

Carlsen, J. and Butler, R. (eds) (2011) *Island Tourism: Sustainable Perspectives*. CABI, Wallingford.

Casaglia, A. (2019) Northern Cyprus as an 'inner neighbour': A critical analysis of European Union enlargement in Cyprus. *European Urban and Regional Studies* 26(1), 37–49.

Caspersen, N. (2011) States without sovereignty: imitating democratic statehood. In: Caspersen, N. and Stansfield, G. (eds) *Unrecognized States in the International System*. Routledge, London, pp. 73–89.

Caspersen, N. (2012) *Unrecognized States: The Struggle for Sovereignty in the Modern International System*. Polity Press, Cambridge.

Caspersen, N. and Stansfield, G. (2011) Introduction: unrecognized states in the International system. In: Caspersen, N. and Stansfield, G. (eds) *Unrecognized States in the International System*. Routledge, London, pp. 1–8.

Castan Pinos, J. and Jagetić Andersen, D. (2012) Challenging the post-Yugoslavian boders: the enclaves of Sastavci and Dubrovnik. In: Jańczak, J. and Osiewicz, P. (eds) *European Exclaves in the Process of De-Bordering and Re-Bordering*. Verlag, Berlin, pp. 81–98.

Catudal, H. (1971) Steinstücken: the politics of a Berlin exclave. *World Affairs* 134(1), 51–62.

Catudal, H. (1974) Exclaves. *Cahiers de géographie du Québec* 18(43), 107–136.

Catudal, H. (1975) The plight of the Lilliputians: an analysis of five European microstates. *Geoforum* 6(3-4), 187–204.

Catudal, H. (1979) *The Exclave Problem of Western Europe*. University of Alabama Press, Tuscaloosa, AL.

Central Intelligence Agency (2020) CIA world Factbook. Available at: https://www.cia.gov/library/publications/the-World-factbook/ (accessed 3 March 2020).

Chacko, H.E. (1996) Positioning a tourism destination to gain a competitive edge. *Asia Pacific Journal of Tourism Research* 1(2), 69–75.

Chaperon, S. (2017) Tourism industry responses to public-private partnership arrangements for destination management organisations in small island economies: a case study of Jersey, Channel Islands. *International Journal of Tourism Policy* 7(1), 23–41.

Chini, M. (2020) Coronavirus: Only half a store opens on Dutch/Belgian border. *The Brussels Times*, 25 March 2020. Available at: https://www.brusselstimes.com/all-news/belgium-all-news/102653/coronavirus-only-half-a-store-opens-on-the-dutch-belgian-border/ (accessed 14 May 2020).

Christiansen, V. and Smith, S. (2008) Optimal commodity taxation with duty-free shopping. *International Tax and Public Finance* 15, 274–296.

Cipollini, C. (2019) The concept of special tax zones in EU tax law. *EC Tax Review* 28(6), 307–321.

Clegg, P. (2016) Brexit and the overseas territories: repercussions for the periphery. *The Round Table* 105(5), 543–555. DOI: 10.1080/00358533.2016.1229420.

Clegg, P. (2018) The United Kingdom and its overseas territories: no longer a 'benevolent patron'? *Small States & Territories* 1(2), 149–168.

Clogg, R. (2015) The Sovereign Base Areas: colonialism redivivus? *Byzantine and Modern Greek Studies* 39(1), 138–150.

Cohen, A. (1993) San Marino, the vest-pocket republic, has big ideas about its place in the world. *Destinations* 8(2), 52–57.

Comas d'Argemir, D. (2002) *Andorra: Una Economía de Frontera*. Pagès, Lleida.

Comas d'Argemir, D. (2010) Tobacco in the Pyrenees? The ecological politics of tobacco in Andorra, a micro-state in a global market. In: Vaccaro, I. and Beltran, O. (eds) *Social and Ecological History of the Pyrenees: State, Market, and Landscape*. Left Coast Press, Walnut Creek, CA, pp. 179–194.

Connell, J. (1988) *Sovereignty & Survival: Island Microstates in the Third World*. Department of Geography, University of Sydney, Sydney.

Connell, J. (1991) Island microstates: the mirage of development. *The Contemporary Pacific* 3(2), 251–287.

Connell, J. (1993) Anguilla: the tourist trajectory in an island microstate. *Caribbean Geography* 4(2), 131–137.

Connell, J. (2007) 'The Best Island on the Globe': constantly constructing tourism on Niue. *The Australian Geographer* 38(1), 1–13.
Connell, J. and Conway, D. (2000) Migration and remittances in island microstates: a comparative perspective on the South Pacific and the Caribbean. *International Journal of Urban and Regional Research* 24(1), 52–78.
Conway, D. (1998) Microstates in a macroworld. In: Klak, T. (ed.) *Globalization and Neoliberalism: The Caribbean Context*. Rowman & Littlefield, Lanham, MD, pp. 51–63.
Cooley, A. and Mitchell, L.A. (2010) Engagement without recognition: a new strategy toward Abkhazia and Eurasia's unrecognized states. *The Washington Quarterly* 33(4), 59–73.
Cooper, C. (1995) Strategic planning for sustainable tourism: the case of the offshore islands of the UK. *Journal of Sustainable Tourism* 3(4), 191–209.
Corbet, H. and Robertson, D. (eds) (2016) *Europe's Free Trade Area Experiment: EFTA and Economic Integration*. Elsevier, Amsterdam.
Corbett, J., Veenendaal, W. and Ugyel, L. (2017) Why monarchy persists in small states: the cases of Tonga, Bhutan and Liechtenstein. *Democratization* 24(4), 689–706.
Cornell, S.E. (1999) *The Nagorno-Karabakh Conflict*. Uppsala University, Uppsala, Sweden.
Council of Europe (2020) Sustainable democratic societies. Available at: https://www.coe.int/en/web/portal (accessed 5 March 2020).
Cox, N. (2006) The continuing question of sovereignty and the Sovereign Military Order of Jerusalem, of Rhodes and of Malta. *Australian International Law Journal* 13, 211–232.
Crowley, N. (2013) *More Armenian than Armenian: Diasporan Tourism Programs and the Rebranding of "Armenia" as Transnation. Unpublished master's thesis*. Central European University, Budapest, Hungary.
Cuevas Contreras, T., Berthet, N. and Urrutia de la Garza, J.A. (2016) Andorra: actual nodal context of tourism between France and Spain. *Transitare: Revista de Turismo, Economía y Negocios* 2(1), 74–99.
Cuevas, T. and Berthet, N. (2012) Andorra, contexto presente del turismo nodal entre Francia y España. *European Scientific Journal* 8(28), 78–101.
Daase, C. (2016) Abchasien, Süd-Ossetien und Transnistrien—de facto-(Klein-)Staaten im postsowjetischen Raum. In: Wolf, S. (ed.) *State Size Matters: Politik und Recht im Kontext von Kleinstaatlichkeit und Monarchie*. Springer, Wiesbaden, pp. 219–249.
Dahlman, C.T. (2009a) Sovereignty. In: Gallaher, C., Dahlman, C.T., Gilmartin, M., Mountz, A. and Shirlow, P. (eds) *Key Concepts in Political Geography*. SAGE, Los Angeles, pp. 28–40.
Dahlman, C.T. (2009b) Territory. In: Gallaher, C., Dahlman, C.T., Gilmartin, M., Mountz, A. and Shirlow, P. (eds) *Key Concepts in Political Geography*. SAGE, Los Angeles, pp. 77–86.
Danmarks Nationalbank (2014) Faroese banknotes. Available at: http://www.nationalbanken.dk/en/banknotes_and_coins/Faroeislands/Pages/Default.aspx (accessed 20 January 2020).
Davies, G. (2017) Making space for the millionaires: Monaco's £1.5billion plan to reclaim 15 acres of land from the Med so thousands can move to the tax haven in the next decade. *Daily Mail*, 20 November 2017. Available at: https://www.dailymail.co.uk/news/article-5099939/Monaco-s-1-5billion-plan-reclaim-land-sea.html (accessed 4 April 2020).
de Albuquerque, K. and McElroy, J.L. (1992) Caribbean small-island tourism styles and sustainable strategies. *Environmental Management* 16(5), 619–632.
de Castro, V.B. and Kober, R. (2018) The Principality of Hutt River: a territory marooned in the Western Australian outback. *Shima* 12(1), 143–159.
de la Haye, M. (2013) Jersey: the states of Jersey. In: Baldwin, N.D.J. (ed.) *Legislatures of Small States: A Comparative Study*. Routledge, London, pp. 116–121.
della Dora, V. (2012) Setting and blurring boundaries: pilgrims, tourists, and landscape in Mount Athos and Meteora. *Annals of Tourism Research* 39(2), 951–974. DOI: 10.1016/j.annals.2011.11.013.
Dellanna, A. and Monella, L.M. (2020) Can San Marino's 'aggressive' COVID-19 testing strategy save Europe? 25 April 2020. Available at: https://www.euronews.com/2020/04/25/can-san-marino-s-aggressive-covid-19-testing-strategy-save-europe (accessed 12 May 2020).
Deppisch, S. (2008) Social capital and other main influences on governance processes in cross-border micro-regions. In: Leibenath, M., Korselli-Olejniczak, E. and Knippschild, R. (eds) *Cross-border Governance and Sustainable Spatial Development: Mind the Gaps!* Springer, Berlin, pp. 67–83.
Dernoi, L.A. (1983) Farm tourism in Europe. *Tourism Management* 4(3), 155–166.

Dharmapala, D. (2008) What problems and opportunities are created by tax havens? *Oxford Review of Economic Policy* 24(4), 661–679.

d'Hauteserre, A.-M. (2001) Bridging the culture nature divide in Monaco. *Landscape and Urban Planning* 57(3-4), 209–223.

d'Hauteserre, A.-M. (2005) Tourism, development and sustainability in Monaco: comparing discourses and practices. *Tourism Geographies* 7(3), 290–312.

Díaz-Sauceda, J., Palau-Saumell, R., Forgas-Coll, S. and Sánchez-García, J. (2015) Cross-Border tourists' behavioral intentions: the green line of Nicosia, Cyprus. *Tourism Geographies* 17(5), 758–779.

Diggines, C.E. (1985) The problems of small states. *The Round Table* 295, 191–205.

Dinica, V. (2012) Tourism on Curaçao: explaining the shortage of sustainability legislation from game theory perspective. *Journal of Environmental Policy & Planning* 14(2), 161–188.

Dodds, K., Jensen, R.B. and Constantinou, C.M. (2015) Signposts: Cyprus, UK and the future of the SBAs. *The RUSI Journal* 160(5), 36–45.

Dommen, E. and Hein, P. (1987) *States, Microstates and Islands*. Croom Helm, London.

Donaldson, J.W. and Williams, A.J. (2008) Delimitation and demarcation: analysing the legacy of Stephen B. Jones's boundary-making. *Geopolitics* 13(4), 676–700.

Dumieński, Z. (2014) *Microstates as Modern Protected States: Towards a New Definition of Micro-Statehood*. Centre for Small States Studies, Reykjavik, Iceland.

Dunford, G. (2009) A mini-history of micronations. *The Futurist* 43(3), 34.

Duursma, J. (1994) *Self-determination, Statehood and International Relations of Micro-States: The Case of Liechtenstein, San Marino, Monaco, Andorra and the Vatican City*. Leiden, University of Leiden.

Duursma, J. (2006) Micro-states: the principality of Liechtenstein. In: Ingebritsen, C., Neumann, I.B. and Gstöhl, S. (eds) *Small States in International Relations*. University of Washington Press, Seattle, pp. 89–147.

Eames, A. (2017) Europe's strange border anomaly. BBC Travel. Available at: http://www.bbc.com/travel/story/20171210-europes-strange-border-anomaly (accessed 7 February 2020).

Eccardt, T.M. (2005) *Secrets of the Seven Smallest States of Europe: Andorra, Liechtenstein, Luxembourg, Malta, Monaco, San Marino, and Vatican City*. Hippocrene Books, New York.

Economist (2020) Goodbye casino, hello smugglers. The Economist. Available at: https://www.economist.com/europe/2020/01/02/a-tiny-italian-exclave-unwillingly-joins-the-eus-customs-union (accessed 22 January 2020).

Edwards, E. (2018) Survey of audience reception in the Sistine Chapel: decoding the message of sacred art. *Church, Communication and Culture* 3(3), 260–282.

EEIG Eifel-Ardennes Marketing (2020) The Vennbahn stories. Available at: https://www.vennbahn.eu/en/stories/ (accessed 27 April 2020).

Egan, M. and Guimarães, M.H. (2017) The single market: trade barriers and trade remedies. *JCMS: Journal of Common Market Studies* 55(2), 294–311.

Eggers, M. (2019) Llívia, a Spanish city landlocked by France. My Magic Earth. March 2. Available at: https://mymagicearth.com/llivia-spanish-city-landlocked-france (accessed 7 February 2020).

Embassy of Liechtenstein (2020) Spirited Liechtenstein. Available at: https://www.liechtensteinusa.org/article/spirited-liechtenstein (accessed 3 March 2020).

Embassy of Monaco (2020) Embassy of Monaco in Washington, DC. Available at: http://monacodc.org/ (accessed 3 May 2020).

Emerson, M. (2007) *Andorra and the European Union*. Centre for European Policy Studies, Brussels.

Entwistle, M. and Oliver, M.J. (2015) Jersey, a small island international finance center: adapting to survive. In: Baldacchino, G. (ed.) *Entrepreneurship in Small Island States and Territories*. Routledge, London, pp. 251–267.

Euro News (2020) Tiny Italian enclave in Switzerland transferred back to Italy and the EU's customs union. Available at: https://www.euronews.com/2020/01/03/tiny-italian-enclave-in-switzerland-transferred-back-to-italy-and-the-eu-s-customs-union (accessed 26 April 2020).

European Commission (2020a) Overseas countries and territories. Available at: https://ec.europa.eu/international-partnerships/where-we-work/overseas-countries-and-territories_en (accessed 3 March 2020).

European Commission (2020b) What is the Economic and Monetary Union? (EMU). Available at: https://ec.europa.eu/info/business-economy-euro/economic-and-fiscal-policy-coordination/economic-and-monetary-union/what-economic-and-monetary-union-emu_en (accessed 19 February 2020).

European Parliament (2020) Fact sheets on the European Union: Outermost Regions (ORs). Available at: https://www.europarl.europa.eu/factsheets/en/sheet/100/outermost-regions-ors- (accessed 30 March 2020).

European Travel Information and Authorization System (2020) ETIAS requirements for Andorran citizens. Available at: https://www.etiasvisa.com/etias-requirements/andorran (accessed 30 March 2020).

Eyres, L. (2020) A tiny 'country' between France and Switzerland. *BBC Travel*, 12 March 2020. Available at: http://www.bbc.com/travel/story/20200311-a-tiny-country-between-france-and-switzerland (accessed 4 May 2020).

Fagence, M. (1997) An uncertain future for tourism in microstates: the case of Nauru. *Tourism Management* 18(6), 385–392.

Fagence, M. (1999) Tourism as a feasible option for sustainable development in small island developing states (SIDS): Nauru as a case study. *Pacific Tourism Review* 3(2), 133–142.

Farran, C.O. (1954) The Sovereign Order of Malta in international law. *International & Comparative Law Quarterly* 3(2), 217–234.

Fehlmann, R., Grahlow, H., Lutz, A., Passow, T. and Schierscher, R. (2002) *Liechtenstein: Introducing the Global Village*. University of Lugano, Lugano, Switzerland.

Fellman, K., Kinnunen, J., Lindström, B. and Palmer, R. (2015) Pathways to successful entrepreneurship in small island economies: the case of Åland. In: Baldacchino, G. (ed.) *Entrepreneurship in Small Island States and Territories*. Routledge, London, pp. 87–102.

Fellman, S. (2016) Sustainable food tourism: a case study on the Åland Islands. Unpublished master's thesis. Mid-Sweden University.

Forsgård, H. (2019) Åland: Many travellers, far fewer overnighters. *Nordic Labour Journal*, 27 June 2019. Available at: http://www.nordiclabourjournal.org/i-fokus/in-focus-2019/sustainable-tourism/article.2019-06-26.4591163555 (accessed 14 February 2020).

Forster, N. and Mallin, F. (2014) *The Association of European Microstates with the EU: Integration Test with Model Value*. Stiftung Wissenschaft und Politik, Deutsches Institut für Internationale Politik und Sicherheit, Berlin.

Forysinski, W. (2012) Exclaves – legal perspectives. In: Jańczak, J. and Osiewicz, P. (eds) *European Exclaves in the Process of De-Bordering and Re-Bordering*. Verlag, Berlin, pp. 21–30.

Fourquet, L. (2017) This Catalan town has already broken from Spain, physically at least. *New York Times*, 24 October 2017. Available at: https://www.nytimes.com/2017/10/24/world/europe/catalonia-independence-spain-llivia.html (accessed 3 December 2019).

Frommelt, C. (2016) Liechtenstein's tailor-made arrangements in the EEA: a small state's creative solutions in European integration. In: Wolf, S. (ed.) *State Size Matters: Politik und Recht im Kontext von Kleinstaatlichkeit und Monarchie*. Springer, Wiesbaden, pp. 131–159.

Føroya Landsstýri (2019) The Faroe Islands in the international community. Available at: https://www.faroeislands.fo/government-politics/foreign-relations/ (accessed 20 March 2020).

Gardner, N. and Kries, S. (2005) Tangled territories. *Hidden Europe* 3, 12–17.

Gay, J.C. (1998) Turning constraints to advantages: the development of tourism in the Principality of Monaco. *Espace Geographique* 27, 169–182.

Gelbman, A. (2010) Border tourism attractions as a space for presenting and symbolizing peace. In: Moufakkir, O. and Kelly, I. (eds) *Tourism, Progress and Peace*. CABI, Wallingford, pp. 83–98.

Gelbman, A. and Timothy, D.J. (2010) From hostile boundaries to tourist attractions. *Current Issues in Tourism* 13(3), 239–259.

Gelbman, A. and Timothy, D.J. (2011) Border complexity, tourism and international exclaves: a case study. *Annals of Tourism Research* 38(1), 110–131.

Gelbman, A. and Timothy, D.J. (2019) Differential tourism zones on the western Canada-US border. *Current Issues in Tourism* 22(6), 682–704.

Gemeinde Büsingen (2020) Büsingen – die deutsche Insel in der Schweiz. Available at: https://www.buesingen.de/willkommen (accessed 1 May 2020).

Gemeinde Jungholz (2020) Gemeinde Jungholz: die Perle Tirols im Oberallgäu. Available at: https://www.jungholz.tirol.gv.at/Unser_Jungholz/Wissenswertes/Zahlen_und_Fakten (accessed 22 April 2020).

German, T. (2016) Russia and South Ossetia: conferring statehood or creeping annexation? *Southeast European and Black Sea Studies* 16(1), 155–167.

Gerrits, A.W. and Bader, M. (2016) Russian patronage over Abkhazia and South Ossetia: implications for conflict resolution. *East European Politics* 32(3), 297–313.

Gilmartin, M. (2009) Nation-state. In: Gallaher, C., Dahlman, C.T., Gilmartin, M., Mountz, A. and Shirlow, P. (eds) *Key Concepts in Political Geography*. SAGE, Los Angeles, pp. 19–27.

Glassner, M. and Fahrer, C. (2004) *Political Geography*, 3rd edn. Wiley, Hoboken, NJ.

Go Vilnius (2020) Užupis. Available at: https://www.govilnius.lt/visit-vilnius/places/uzupis-district (accessed 30 April 2020).

Goetschel, L. (1998) The foreign and security policy interests of small states in today's Europe. In: Goetschel, L. (ed.) *Small States Inside and Outside the European Union: Interests and Policies*. Kluwer, Dortrecht, pp. 13–31.

Gold, P. (2000) *Europe or Africa? A Contemporary Study of the Spanish North African Enclaves of Ceuta and Melilla*. Oxford University Press, Oxford.

Gold, P. (2005) *Gibraltar: British or Spanish?* Routledge, London.

Gosh, B.N. (2019) *Dependency Theory Revisited*. Taylor & Francis, London.

Government of Monaco (2020) *Monaco en Chiffres: Économie*. Monaco-Ville, Government of Monaco.

Graubünden Ferien (2020) Duty free shopping in Samnaun. Available at: https://www.graubuenden.ch/en/discover-our-regions/scuol-samnaun-val-mustair/duty-free-shopping-samnaun (accessed 27 April 2020).

Grimaldi, A.I. (2006) Monaco and marine environmental protection. *IAEA Bulletin* 48(1), 61.

Grinda, G. (2006) *The Principality of Monaco: State, International Status, Institutions*. Asser Press, The Hague.

Grydehøj, A. (2010) Uninherited heritage: tradition and heritage production in Shetland, Åland and Svalbard. *International Journal of Heritage Studies* 16(1-2), 77–89.

Grydehøj, A. (2011) Making the most of smallness: economic policy in microstates and sub-national island jurisdictions. *Space and Polity* 15(3), 183–196.

Grydehøj, A. (2018) Decolonising the economy in micropolities: rents, government spending and infrastructure development in Kalaallit Nunaat (Greenland). *Small States & Territories* 1(1), 69–94.

Grydehøj, A., Grydehøj, A. and Ackrén, M. (2012) The globalization of the Arctic: negotiating sovereignty and building communities in Svalbard, Norway. *Island Studies Journal* 7(1), 99–118.

Guiberteau, O. (2019) The British island that's not in the UK. BBC Travel. Available at: http://www.bbc.com/travel/story/20191008-the-british-isle-that-doesnt-belong (accessed 15 October 2019).

Gumba, Y. and Ketsba, T. (2004) Economic development prospects in Abkhazia and the concept of regional cooperation. In: Champain, P., Klein, D. and Mirimanova, N. (eds) *From War Economies tot Peace Economies in the South Caucasus*. International Alert, London, pp. 158–176.

Gunkel, C. (2014) Bargain-hunting tourists flock to Heligoland. DW.com. Available at: https://www.dw.com/en/bargain-hunting-tourists-flock-to-heligoland/a-17792274 (accessed 4 May 2020).

Guo, Y. and Wu, J.J. (eds) (2016) *Singapore and Switzerland: Secrets to Small State Success*. World Scientific, Singapore.

Hall, C.M. (1994a) Tourism in Pacific island microstates: a case study of Vanuatu. *Tourism Recreation Research* 19(1), 59–64.

Hall, C.M. (1994b) *Tourism in the Pacific Rim: Development, Impacts, and Markets*. Longman, London.

Hall, C.M. (2014) Regulating the international trade in tourism services. In: Coles, T. and Hall, C.M. (eds) *International Business and Tourism: Global Issues, Contemporary Interactions*. Routledge, London, pp. 47–68.

Hall, C.M. (2015) On the mobility of tourism mobilities. *Current Issues in Tourism* 18(1), 7–10.

Hall, C.M. (2017) Tourism and geopolitics: the political imaginary of territory, tourism and space. In: Hall, D. (ed.) *Tourism and Geopolitics: Issues and Concepts from Central and Eastern Europe*. CABI, Wallingford, pp. 15–24.

Hall, D. (ed.) (2004) *Tourism in Transition: Governance, Transformation and Development*. CABI, Wallingford.

Hall, D. (2017a) Bringing geopolitics to tourism. In: Hall, D. (ed.) *Tourism and Geopolitics: Issues and Concepts from Central and Eastern Europe*. CABI, Wallingford, pp. 3–14.

Hall, D. (ed.) (2017b) *Tourism and Geopolitics: Issues and Concepts from Central and Eastern Europe*. CABI, Wallingford.

Hall, D. (2020) *Brexit and Tourism: Process, Impacts and Non-Policy*. Channel View Publications, Bristol.

Hamilton, J.A. (1923) The Republic of San Marino. *The Scottish Geographical Magazine* 39(4), 243–251.

Hampton, M.P. and Christensen, J. (2007) Competing industries in islands: a new tourism approach. *Annals of Tourism Research* 34(4), 998–1020.

Hannam, K. (2013) "Shangri-La" and the new "great game": Exploring tourism geopolitics between China and India. *Tourism Planning & Development* 10(2), 178–186.

Harrison, D. (1996) Sustainability and tourism: reflections from a muddy pool. In: Briguglio, L., Archer, B., Jafari, J. and Wall, G. (eds) *Sustainable Tourism in Islands and Small States: Issues and Policies*. Pinter, London, pp. 69–89.
Harrison, D. (2001) Tourism in small islands and microstates. *Tourism Recreation Research* 26(3), 3–8.
Harrison, D. (2004) Tourism in Pacific Islands. *The Journal of Pacific Studies* 26(1/2), 1–28.
Hartshorn, R. (1936) Suggestions on the terminology of political boundaries. *Annals of the Association of American Geographers* 26, 56–57.
Hayat, D.T. (2019) Archaeological heritage management in Turkish Republic of Northern Cyprus: implication for sustainable tourism. *Journal of Tourism and Hospitality Management* 7(1), 127–134.
Hendry, I. and Dickson, S. (2018) *British Overseas Territories Law*. Bloomsbury, London.
Hepburn, E. and Baldacchino, G. (eds) (2016) *Independence Movements in Subnational Island Jurisdictions*. Routledge, London.
Hoch, T. (2020a) De facto statehood: overview of the research. In: Hoch, T. and Kopeček, V. (eds) *De Facto States in Eurasia*. Routledge, London, pp. 16–26.
Hoch, T. (2020b) Factors of de facto states' formation in the post-Soviet area. In: Hoch, T. and Kopeček, V. (eds) *De Facto States in Eurasia*. Routledge, London, pp. 81–85.
Hoch, T. and Souleimanov, E.A. (2020) Formation of de facto states in Abkhazia and South Ossetia. In: Hoch, T. and Kopeček, V. (eds) *De Facto States in Eurasia*. Routledge, London, pp. 86–108.
Hoffmann, A.M. (2016) (Un-)Abhängige Staatlichkeit? Der Mikrostaat als internationaler Akteur: Kleinstaatlichkeit und das Entscheidungsverhalten ostkaribischer Inselstaaten in der internationalen Politik. In: Wolf, S. (ed.) *State Size Matters: Politik und Recht im Kontext von Kleinstaatlichkeit und Monarchie*. Springer, Wiesbaden, pp. 251–275.
Holmberg-Anttila, E. (2009) Tourism in the Åland Islands: an overview. In: Hall, C.M., Müller, D.K. and Saarinen, J. (eds) *Nordic Tourism: Issues and Cases*. Channel View Publications, Bristol, pp. 111–114.
House of Commons Library (2020) How autonomous are the crown dependencies? Available at: https://commonslibrary.parliament.uk/world-affairs/institutions/%EF%BB%BFhow-autonomous-are-the-crown-dependencies/ (accessed 29 April 2020).
Hovgaard, G. and Ackrén, M. (2017) *Autonomy in Denmark: Greenland and the Faroe Islands*. Barcelona, Centre for International Affairs.
Il Resto del Carlino (2012) L'idea: Ripristinare la ferrovia tra Rimini e San Marino. Available at: https://www.ilrestodelcarlino.it/rimini/cronaca/2012/06/28/736096-ferrovia-rimini-san-marino.shtml (accessed 22 March 2020).
Ingebritsen, C., Neumann, I.B. and Gstöhl, S. (eds) (2006) *Small States in International Relations*. University of Washington Press, Seattle.
Instituto Nazionale di Statistica (2019) *Movimento Turistico in Italia, Anno 2018*. ISTAT, Rome.
International Business Publications (2013) *Liechtenstein Investment and Business Guide*. International Business Publications, Washington, DC.
International Business Publications (2017) *Liechtenstein Country Study Guide*. International Business Publications, Washington, DC.
International Business Publications (2019) *San Marino: Doing Business and Investing in San Marino Guide*. International Business Publications, Washington, DC.
Ivanov, S., Idzhylova, K. and Webster, C. (2016) Impacts of the entry of the autonomous Republic of Crimea into the Russian Federation on its tourism industry: an exploratory study. *Tourism Management* 54, 162–169.
Jacobson, D., Webster, C., Shapiro, K., Musyck, B. and Orphanides, S. (2015) Cyprus settlement: a zero sum game for tourism? *European Journal of Tourism Research* 11, 21–34.
Jahn, E. (2015) Linguistic assimilation of all citizens or minority protection: the precedent set by the Åland Islands. In: Jahn, E. (ed.) *World Political Challenges: Political Issues Under Debate—Volume 3*. Springer, Berlin, pp. 239–256.
Jakola, F. and Prokkola, E.-K. (2018) Trust building or vested interest? Social capital processes of cross-border co-operation in the border towns of Tornio and Haparanda. *Tijdschrift voor Economische en Sociale Geografie* 109(2), 224–238.
Jansson, D. (2012) Branding Åland, branding Ålanders: reflections on place identity and globalization in a Nordic Archipelago. *Place Branding and Public Diplomacy* 8(2), 119–132.
Jańczak, J. (2012) Baarle-Hertog and Baarle-Nassau: functional interdependence of the nested territorial and political structures. In: Jańczak, J. and Osiewicz, P. (eds) *European Exclaves in the Process of De-Bordering and Re-Bordering*. Verlag, Berlin, pp. 57–70.

Jenner, P. and Smith, C. (1993) Europe's micro-states. *International Tourism Reports* 1, 68–89.
Jersey Heritage (2020) Bergerac's Island. Available at: https://www.jerseyheritage.tv/bergerac-s-island (accessed 1 May 2020).
Johansson, R. (1991) Farm tourism on the Aland archipelago. *Tourism Recreation Research* 16(1), 62–63.
Jones, S. (1945) *Boundary Making*. Carnegie Endowment, Washington, DC.
Jones, S. (1959) Boundary concepts in setting time and space. *Annals of the Association of American Geographers* 49, 241–155.
Juanola, M.P. (2020) Hutt River Principality will close its doors 'until further notice' due to financial hardship. WAtoday. Available at: https://www.watoday.com.au/national/western-australia/hutt-river-principality-will-close-its-doors-until-further-notice-due-to-financial-hardship-20200110-p53qja.html (accessed 30 April, 2020).
Kalicka-Mikołajczyk, A. (2018) Cooperation between the European Union and European mini-states–Possible scenarios for future closer and enhanced integration with the European Union. *Wroclaw Review of Law, Administration & Economics* 8(1), 56–66.
Karlsson, A. (2006) *Sub-National Insular Jurisdictions as Configurations of Jurisdictional Powers and Economic Capacity: A Study of Åland, the Faroe Islands and Greenland*. Nordregio, Stockholm.
Karlsson, A. (2009) Sub-national island jurisdictions as configurations of jurisdictional powers and economic capacity: Nordic experiences from Åland, Faroes and Greenland. *Island Studies Journal* 4(2), 139–162.
Karlsson, T. (1999) A tax paradise in the making? Alcohol regulations in the Åland Islands. *Contemporary Drug Problems* 26(1), 3–30.
Katnic, M. and Boskovic, B. (2019) The competitiveness of small European countries: a focus on Montenegro. *Small States & Territories* 2(1), 9–22.
Keller, W. and Shiue, C.H. (2014) Endogenous formation of free trade agreements: evidence from the Zollverein's impact on market integration. *Journal of Economic History* 74(4), 1168–1204.
Kelman, I. and Shreve, C. (2015) Dark entrepreneurship in small island states and territories. In: Baldacchino, G. (ed.) *Entrepreneurship in Small Island States and Territories*. Routledge, London, pp. 31–47.
Kentas, G. (2018) A critical assessment of the Cyprus Protocol annexed to the UK's withdrawal agreement. *The Cyprus Review* 30(1), 317–348.
Khan, D. (2004) *Die deutschen Staatsgrenzen*. Mohr Siebeck, Tübingen.
Kim, B. and Kim, S. (2019) Hierarchical value map of religious tourists visiting the Vatican City/Rome. *Tourism Geographies* 21(3), 529–550.
King, D.C. (2008) *Cultures of the World: Monaco*. Marshall Cavendish, New York.
King, J. (2013) The Isle of Man: Tynwald. In: Baldwin, N.D.J. (ed.) *Legislatures of Small States: A Comparative Study*. Routledge, London, pp. 122–133.
Kleinwalsertal Tourismus (2020) Dreaming of a mountain summer? So are we. Available at: https://www.kleinwalsertal.com/en (accessed 29 April, 2020).
Klemenčić, M. and Gosar, A. (2000) The problems of the Italo-Croato-Slovene border delimitation in the Northern Adriatic. *GeoJournal* 52(2), 129–137.
Klieger, P.C. (2013) *The Microstates of Europe: Designer Nations in a Post-modern World*. Lexington Books, Lanham, MD.
Kliot, N. and Mansfeld, Y. (1994) Resettling displaced people in North and South Cyprus: a comparison. *Journal of Refugee Studies* 7(4), 328–359.
Kliot, N. and Mansfeld, Y. (1997) The political landscape of partition: the case of Cyprus. *Political Geography* 16(6), 495–521.
KNOEMA (2020) Transportation: road density. Available at: https://knoema.com/atlas/ranks/Road-density (accessed 15 March 2020).
Kochenov, D. (ed.) (2011) *EU Law of the Overseas: Outermost Regions, Associated Overseas Countries and Territories, Territories Sui Generis*. Kluwer Law, Alphen aan den Rijn.
Kolstø, P. (2020) Biting the hand that feeds them? Abkhazia-Russia client-patron relations. *Post-Soviet Affairs* 36(2), 140–158.
Kolstø, P. and Blakkisrud, H. (2012) De facto states and democracy: the case of Nagorno-Karabakh. *Communist and Post-Communist Studies* 45(1-2), 141–151.
Kopeček, V. (2020a) De facto states and other unrecognized entities in Eurasia. In: Hoch, T. and Kopeček, V. (eds) *De Facto States in Eurasia*. Routledge, London, pp. 27–39.

Kopeček, V. (2020b) Nagorno-Karabakh and Javakheti: two different trajectories of Armenian separatist movements. In: Hoch, T. and Kopeček, V. (eds) *De Facto States in Eurasia*. Routledge, London, pp. 208–224.

Kosienkowski, M. (2020) The patron-client relationship between Russia and Transnistria. In: Hoch, T. and Kopeček, V. (eds) *De Facto States in Eurasia*. Routledge, London, pp. 183–207.

Kotsi, F. (2012) Mount Athos: development policies for short-term religious tourism. *International Journal of Tourism Anthropology* 2(2), 149–163.

Kovova, I., Malyshkin, O., Vicen, V., Shulyarenko, S., Semenova, S. *et al*. (2018) Value added tax: effectiveness and legal regulation in Ukraine and the European Union. *Economic Annals* 171(5-6), 4–14.

Krogh, J. (2012) Vennbahn: German exclaves in Belgium. Available at: http://geosite.jankrogh.com/vennbahn.htm (accessed 3 November 2019).

Kwiatkiewicz, P. (2012) Nakhchivan autonomous Republic and Kerki, Bakhudarli, Yukhary – Askipara. In: Jańczak, J. and Osiewicz, P. (eds) *European Exclaves in the Process of De-Bordering and Re-Bordering*. Verlag, Berlin, pp. 99–111.

Kyris, G. (2020) The European Union in Northern Cyprus: Conceptualising the avoidance of contested states. *Geopolitics* 25(2), 346–361.

Lambert, M.E. (2018) The foreign policy option of Western countries regarding de facto states in the post-Soviet space. In: Labarre, F. and Niculescu, G. (eds) *What a 'New European Security Deal' Could Mean for the South Caucasus*. Republic of Austria, Federal Ministry of Defense, Vienna, pp. 111–123.

Lavenex, S. (2011) Concentric circles of flexible 'EUropean' integration: a typology of EU external governance relations. *Comparative European Politics* 9(4-5), 372–393.

Lea, J. (2006) *Tourism and Development in the Third World*. Routledge, London.

Leadbeater, C. (2019) Discover the Channel Island that inspired a literary masterpiece. The Telegraph. Available at: https://www.telegraph.co.uk/travel/discover-guernsey/inspiration-for-les-miserables/ (accessed 10 November, 2019).

Leigh, J. and Vukovic, P. (2011) A geopolitics of Cyprus. *Middle East Review of International Affairs* 15(4), 59–70.

Leiper, N. (1989) Tourism and gambling. *GeoJournal* 19(3), 269–275.

Lew, A.A., Hall, C.M. and Timothy, D.J. (2015) *World Regional Geography: Human Mobilities, Tourism Destinations, Sustainable Environments*, 2nd edn. Kendall-Hunt, Dubuque, Iowa.

Liechtenstein Tourism (2005) *Liechtenstein, Vacationland*. Liechtenstein Tourism, Vaduz.

Liguria, R. (2017) *Ancienne Principauté. Seborga*. Azienda Speciale della Camera di Commercio di Imperia Riviera dei Fiori, Seborga.

Lindström, B. (2000) Culture and economic development in Åland. In: Baldacchino, G. and Milne, D. (eds) *Lessons from the Political Economy of Small Islands*. Palgrave Macmillan, London, pp. 107–121.

Liu, Z.-H. and Jenkins, C.L. (1996) Country size and tourism development: a cross-nation analysis. In: Briguglio, L., Archer, B., Jafari, J. and Wall, G. (eds) *Sustainable Tourism in Islands and Small States: Issues and Policies*. Pinter, London, pp. 90–117.

Liutikas, D. (2018) Catholic pilgrimage in Europe: contemporary traditions and challenges. In: El-Gohary, H., Edwards, D.J. and Eid, R. (eds) *Global Perspectives on Religious Tourism and Pilgrimage*. IGI Global, Hershey, PA, pp. 84–103.

Lockhart, D., Schembri, P.J. and Smith, D.W. (eds) (2002) *The Development Process in Small Island States*. Routledge, London.

Loughlin, J. and Daftary, F. (1999) *Insular Regions and European Integration: Corsica and the Åland Islands Compared*. European Centre for Minority Issues, Flensburg, Germany.

Łukaszewski, M. (2015) Microstate and monarchy in the face of the challenges of the modern world: the political system of Liechtenstein and its specificity (an outline of the problem). *Przegląd Politologiczny* 3, 93–102.

Lundén, T. (2012) Exclaves – geographical and historical perspectives. In: Jańczak, J. and Osiewicz, P. (eds) *European Exclaves in the Process of De-Bordering and Re-Bordering*. Verlag, Berlin, pp. 11–20.

Lyon, R. (2010) The story behind a map: the territory of Monaco before 1860. *IMCoS Journal* 123, 34–35.

Lyubitseva, O.A. and Zavarika, G.M. (2019) Geographic analysis of the condition of tourism of Luhansk region (Ukraine) in the context of the conflict. *Journal of Geology, Geography and Geoecology* 28(4), 692–705.

Maass, M. (2017) *Small States in World Politics: The Story of Small State Survival, 1648-2016*. Manchester University Press, Manchester.

MacFarquhar, N. (2018) Mount Athos, a male-only holy retreat, is ruffled by tourists and Russia. *New York Times*, 20 October 2018. Available at: https://www.nytimes.com/2018/10/20/world/europe/mount-athos-greece-russia-eastern-orthodox-church.html (accessed 4 May 2020).

MacLean, R. (2013) Heroic adventures in Transnistria. *New Eastern Europe* 9(4), 152–161.

Mäkelä, S. (2009) Cottage holidays in Åland. In: Hall, C.M., Müller, D.K. and Saarinen, J. (eds) *Nordic Tourism: Issues and Cases*. Channel View Publications, Bristol, pp. 190–191.

Mansfeld, Y. and Kliot, N. (1996) The tourism industry in the partitioned island of Cyprus. In: Pizam, A. and Mansfeld, Y. (eds) *Tourism, Crime and International Security Issues*. Wiley, Chichester, pp. 187–202.

McDowell, B. (1991) *Inside the Vatican*. National Geographic Society, Washington, DC.

McMahon, R. (2013) Guernsey: the states of deliberation of Guernsey. In: Baldwin, N.D.J. (ed.) *Legislatures of Small States: A Comparative Study*. Routledge, London, pp. 107–115.

Mehmet, O. (2010) *Sustainability of Microstates: The Case of North Cyprus*. University of Utah Press, Salt Lake City.

Mehmet, O. and Tahiroglu, M. (2002) Growth and equity in microstates: does size matter in development? *International Journal of Social Economics* 29(1/2), 152–162.

Mehmet, O. and Tahiroglu, M. (2003) Globalization and sustainability of small states. *Humanomics* 19(1), 45–59.

Metreveli, M. and Timothy, D.J. (2010) Effects of the August 2008 war in Georgia on tourism and its resources. In: Moufakkir, O. and Kelly, I. (eds) *Tourism, Progress and Peace*. CABI, Wallingford, pp. 134–147.

Middleton, N. (2015) *An Atlas of Countries That Don't Exist: A Compendium of Fifty Unrecognized and Largely Unnoticed States*. Macmillan, London.

Miles, K. (2019) 'Uneven Empires': Extraterritoriality and the early trading companies. In: Margolies, D.S., Özsu, U., Pal, M. and Tzouvala, N. (eds) *The Extraterritoriality of Law: History, Theory, Politics*. Routledge, London, pp. 87–103.

Milne, S. (1990) The impact of tourism development in small Pacific island states: an overview. *New Zealand Journal of Geography* 89(1), 16–21. DOI: 10.1111/j.0028-8292.1990.tb00283.x.

Milne, S. (1992) Tourism and development in South Pacific microstates. *Annals of Tourism Research* 19(2), 191–212. DOI: 10.1016/0160-7383(92)90077-3.

Misra, A. (2004) Theorising 'small' and 'micro' state behaviour using the Maldives, Bhutan and Nepal. *Contemporary South Asia* 13(2), 133–148. DOI: 10.1080/0958493042000242927.

Mkrtchyan, L. (2012) The exclave of Artshvashen. In: Jańczak, J. and Osiewicz, P. (eds) *European Exclaves in the Process of De-Bordering and Re-Bordering*. Verlag, Berlin, pp. 113–122.

Monaco Government Tourist Office (2012) *Monaco*. Monaco Government Tourist Office, Monte Carlo.

Monaco Tourist and Convention Authority (2020) Visit Monaco: there is no place like here. Available at: https://www.visitmonaco.com/en (accessed 3 April 2020).

Moore, R.F. (1960) *Principality of Liechtenstein: A Brief History*. Simmons-Boardman, New York.

Morris, S.B. (2019) How much does the Vatican make from tourism? Pope2you.net, 20 June. Available at: https://www.pope2you.net/how-much-does-the-vatican-make-from-tourism/ (accessed 19 February 2020).

Mostafanezhad, M. and Norum, R. (2016) Towards a geopolitics of tourism. *Annals of Tourism Research* 61, 226–228. DOI: 10.1016/j.annals.2016.08.003.

Muller, D. (1977) The situation of tourism on the island of Heligoland. *Zeitschrift fur Wirtschaftsgeographie* 21(8), 239–243.

Munar, A.M. and Jacobsen, J.K.S. (2014) Motivations for sharing tourism experiences through social media. *Tourism Management* 43, 46–54. DOI: 10.1016/j.tourman.2014.01.012.

Municipality of Baarle-Hertog (2020) Collaboration with Baarle-Nassau. Available at: https://www.baarle-hertog.be/samenwerking-met-baarle-nassau (accessed 25 April 2020).

Municipality of Campione d'Italia (2020) Information. Available at: http://www.comune.campione-d-italia.co.it/ (accessed 30 March 2020).

Muñoz, J.J. (2017) Intercultural Europe: cultural diversity in the EU and the debate on a common European cultural identity. *Revista de Estudios Europeos* 69, 45–57.

Murray, F. (2012) *The European Union and Member State Territories: A New Legal Framework Under the EU Treaties*. TMC Asser Press, The Hague.

Musiał-Karg, M. (2012) Switzerland's German and Italian islands – Büsingen and Campione d'Italia. In: Jańczak, J. and Osiewicz, P. (eds) *European Exclaves in the Process of De-Bordering and Re-Bordering*. Verlag, Berlin, pp. 31–43.

Mut Bosque, M. (2012) The Spanish exclave of Llívia. In: Jańczak, J. and Osiewicz, P. (eds) *European Exclaves in the Process of De-Bordering and Re-Bordering*. Verlag, Berlin, pp. 45–56.

Mut Bosque, M. (2017) Brexit and the commonwealth: new challenges for Gibraltar. *The Round Table* 106(4), 483–485. DOI: 10.1080/00358533.2017.1352154.

Naurin, D. and Lindahl, R. (2010) Out in the cold? flexible integration and the political status of Euro opt-outs. *European Union Politics* 11(4), 485–509. DOI: 10.1177/1465116510382463.

Neumann, I.B. and Gstöhl, S. (2006) Introduction: Lilliputians in Gulliver's World? In: Ingebritsen, C., Neumann, I.B. and Gstöhl, S. (eds) *Small States in International Relations*. University of Washington Press, Seattle, pp. 3–37.

Newman, D. (2006) The lines that continue to separate us: borders in our `borderless' world. *Progress in Human Geography* 30(2), 143–161. DOI: 10.1191/0309132506ph599xx.

Nielsen, J.K. (2001) Government, culture and sustainability in Greenland: a microstate with a hinterland. *Public Organization Review* 1(2), 229–243. DOI: 10.1023/A:1011504628688.

Niikawa, S. and Corcaci, A. (2016) Reconsidering coalition theory: a set theoretical assessment of government formation in Liechtenstein and Malta. In: Wolf, S. (ed.) *State Size Matters: Politik und Recht im Kontext von Kleinstaatlichkeit und Monarchie*. Springer, Wiesbaden, pp. 33–58.

Nikel, D. (2020) Norway authorities to crack down on Svalbard tour guides. Life in Norway, 16 January 2020. Available at: https://www.lifeinnorway.net/svalbard-tour-guide-crackdown/ (accessed 3 April 2020).

Nilsson, J.H., Eskilsson, L. and Ek, R. (2010) Creating cross-border destinations: Interreg programmes and regionalisation in the Baltic Sea area. *Scandinavian Journal of Hospitality and Tourism* 10(2), 153–172. DOI: 10.1080/15022250903561978.

Nolan, M.L. and Nolan, S. (1992) Religious sites as tourism attractions in Europe. *Annals of Tourism Research* 19(1), 68–78. DOI: 10.1016/0160-7383(92)90107-Z.

Office of Statistics (2019) *Liechtenstein in Figures, 2019*. Office of Statistics, Vaduz.

O'Kane, P. (2014) A history of the 'triple crown' of motor racing: the Indianapolis 500, the Le Mans 24 Hours and the Monaco Grand Prix. In: Hassan, D. (ed.) *The History of Motor Sport: A Case Study Analysis*. Routledge, London, pp. 110–128.

Olafsson, Á. (2000) Constitutionalism and economics in the Faroes. In: Baldacchino, G. and Milne, D. (eds) *Lessons from the Political Economy of Small Islands: The Resourcefulness of Jurisdiction*. Palgrave Macmillan, London, pp. 121–140.

Oliver, M.J. (2019) A small island territory moving down the 'development ladder'? A case of Jersey. *Small States & Territories* 2(1), 83–104.

O'Loughlin, J., Kolossov, V. Gearóid, T. and Toal, G. (2011) Inside Abkhazia: survey of attitudes in a *de facto* state. *Post-Soviet Affairs* 27(1), 1–36. DOI: 10.2747/1060-586X.27.1.1.

O'Loughlin, J., Kolossov, V. and Toal, G. (2014) Inside the post-Soviet de facto states: a comparison of attitudes in Abkhazia, Nagorny Karabakh, South Ossetia, and Transnistria. *Eurasian Geography and Economics* 55(5), 423–456. DOI: 10.1080/15387216.2015.1012644.

O'Reilly, G. (1999) Gibraltar: sovereignty disputes and territorial waters. *IBRU Boundary and Security Bulletin* 7(1), 67–81.

Osiewicz, P. (2012) The Cypriot exclaves: Ormidhia, Xylotymbou and Dhekelia Power Station. In: Jańczak, J. and Osiewicz, P. (eds) *European Exclaves in the Process of De-Bordering and Re-Bordering*. Verlag, Berlin, pp. 21–30.

Ostergren, R. (2010) Defining Liechtenstein: sovereign borders, offshore banking and national identity. In: Diener, A.C. and Hagen, J. (eds) *Borderlines and Borderlands: Political Oddities at the Edge of the Nation-State*. Rowman & Littlefield, Lanham, MD, pp. 137–154.

Page, S. and Connell, J. (2006) *Tourism: A Modern Synthesis*, 2nd edn. Thomson, London.

Paldam, M. (2013) Safe havens in Europe: Switzerland and the ten dwarfs. *The European Journal of Comparative Economics* 10(3), 377–396.

Panke, D. (2011) Microstates in negotiations beyond the nation-state: Malta, Cyprus and Luxembourg as active and successful policy shapers? *International Negotiation* 16(2), 297–317. DOI: 10.1163/138234011X573057.

Panzeri, L. (2019) Autonomia territoriale e specificit di un Comune exclavé: Campione d'Italia. *Le Regioni* 47(2), 571–580.

Parrish, A. (2019) The interplay between extraterritoriality, sovereignty, and the foundations of international law. In: Margolies, D.S., Özsu, U., Pal, M. and Tzouvala, N. (eds) *The Extraterritoriality of Law: History, Theory, Politics*. Routledge, London, pp. 169–182.

Passow, T., Fehlmann, R. and Grahlow, H. (2005) Country reputation—From measurement to management: the case of Liechtenstein. *Corporate Reputation Review* 7(4), 309–326. DOI: 10.1057/palgrave.crr.1540229.

Patawary, K. (2012) The curious case of Baarle-Nassau and Baarle-Hertog. Amusing Planet, 6 November, 2012. Available at: https://www.amusingplanet.com/2012/11/the-curious-case-of-baarle-nassau-and.html (accessed 3 May 2020).

Plieninger, T., Rana, Halla Áargarð av., Fagerholm, N., Ellingsgaard, G.F., Magnussen, E. *et al.* (2018) Identifying and assessing the potential for conflict between landscape values and development preferences on the Faroe Islands. *Global Environmental Change* 52, 162–180. DOI: 10.1016/j.gloenvcha.2018.07.006.

Plog, S. (2001) Why destination areas rise and fall in popularity: an update of a Cornell Quarterly classic. *The Cornell Hotel and Restaurant Administration Quarterly* 42(3), 13–24. DOI: 10.1016/S0010-8804(01)81020-X.

Pons-Pons, M., Johnson, P.A., Rosas-Casals, M., Sureda, B. and Jover, È. (2012) Modeling climate change effects on winter ski tourism in Andorra. *Climate Research* 54(3), 197–207. DOI: 10.3354/cr01117.

Popescu, N. (2006a) *Democracy in Secessionism: Transnistria and Abkhazia's Domestic Policies*. International Policy Fellowship Program, Open Society Institute, Budapest.

Popescu, N. (2006b) 'Outsourcing' de facto statehood: Russia and the secessionist entities in Georgia and Moldova. *CEPS Policy Briefs* 1(12), 1–8.

Population Data (2020) Populationdata.net. Available at: https://en.populationdata.net/countres (accessed 1 April 2020).

Prati, A., McHugh, J. and de Gyor, P.G. (1996) *San Marino, Recent Economic Developments*. International Monetary Fund, Washington, DC.

Pratt, S. (2015) The economic impact of tourism in SIDS. *Annals of Tourism Research* 52, 148–160. DOI: 10.1016/j.annals.2015.03.005.

Prideaux, B. (2008) *Faroe Islands*. In: Lück, Michael, M. (ed.) *The Encyclopedia of Tourism and Recreation in Marine Environments, p. 171*. CABI, Wallingford.

Prokkola, E.-K. (2008) Resources and barriers in tourism development: cross-border cooperation, regionalization and destination building at the Finnish-Swedish border. *Fennia* 186(1), 31–46.

Prokkola, E.-K. (2011) Regionalization, tourism development and partnership: the European Union's North Calotte Sub-programme of INTERREG III a North. *Tourism Geographies* 13(4), 507–530. DOI: 10.1080/14616688.2011.570371.

Raton, P. (1958) Les enclaves. *Annuaire Français de Droit International* 4(1), 186–195. DOI: 10.3406/afdi.1958.1373.

Raton, P. (1970) *Liechtenstein: History and Institutions of the Principality*. Liechtenstein-Verlag, Vaduz.

Redazione ANSA (2016) Pope Francis to receive Knights of Malta grand master Thursday. *Redazione ANSA*, 21 June, 2016. Available at: https://www.ansa.it/english/news/2016/06/21/pope-to-receive-knights-of-malta-grand-master-thursday_a9e72a37-ec6c-421d-a632-92f2dfe81dd7.html (accessed 3 March 2020).

Reichart, T. (1988) Socio-economic difficulties in developing tourism in small alpine countries: the case of Andorra. *Tourism Recreation Research* 13(1), 27–32. DOI: 10.1080/02508281.1988.11014521.

Relph, E. (1976) *Place and Placelessness*. Pion, London.

Reynolds, S. (1996) *Fiefs and Vassals: The Medieval Evidence Reinterpreted*. Oxford University Press, Oxford.

Rhone, E. (2018) Užupis: a tiny Republic of free spirits. BBC Travel. Available at: http://www.bbc.com/travel/story/20181014-uupis-a-tiny-republic-of-free-spirits (accessed 3 May 2020).

Rich, T.S. (2014) Integrating microstates into cross-national research: an exploratory analysis. *Journal of International and Global Studies* 6(1), 1–13.

Riezman, R. (2013) Customs unions and the core. In: Riezman, R. (ed.) *International Trade Agreements and Political Economy*. World Scientific, Hackensack, NJ, pp. 33–43.

Rinschede, G. (1977) Andorra-from secluded high mountain state to international tourist centre. *Erdkunde* 31(4), 307–314.

Robinson, E. (ed.) (1960) *Economic Consequences of the Size of Nations*. Macmillan, London.

Robinson, G.W.S. (1953) West Berlin: the geography of an exclave. *Geographical Review* 43(4), 540–557. DOI: 10.2307/212038.

Robinson, G.W.S. (1958) Ceuta and Melilla: Spain's Plazas de Soberanía. *Geography* 43(4), 266–269.

Robinson, G.W.S. (1959) Exclaves. *Annals of the Association of American Geographers* 49(3), 283–295. DOI: 10.1111/j.1467-8306.1959.tb01614.x.

Ron, A.S. and Timothy, D.J. (2019) *Contemporary Christian Travel: Pilgrimage, Practice and Place*. Channel View Publications, Bristol.

Rosenthal, D.E. and Knighton, W.M. (2017) *National Laws and International Commerce: The Problem of Extraterritoriality*. Routledge, London.

Rougier, H. and Sanguin, A.-L. (1981) Deux zones franches au cœur des Alpes Centrales : Livigno et Samnaun. *Revue de Géographie Alpine* 69(4), 543–560. DOI: 10.3406/rga.1981.2479.

Ryan, C. (2001) Tourism in the South Pacific—A case of marginalities. *Tourism Recreation Research* 26(3), 43–49. DOI: 10.1080/02508281.2001.11081198.

Ryan, J., Dunford, G. and Sellars, S. (2006) *Micronations*. Lonely Planet, Hawthorne, Australia.

Sahlins, P. (1989) *Boundaries: The Making of France and Spain in the Pyrenees*. University of California Press, Berkeley.

Sánchez Arsuaga, G. (2018) *The Pheasant Island: A Threshold in Time*. Retroactive Research, Alicante.

Schratzenstaller, M. and Krenek, A. (2019) Tax-based own resources to finance the EU budget. *Intereconomics* 54(3), 171–177. DOI: 10.1007/s10272-019-0817-0.

Seddighi, H.R., Theocharous, A.L. and Nuttall, M.W. (2002) Political instability and tourism: an empirical study with special reference to the microstate of Cyprus. *International Journal of Hospitality & Tourism Administration* 3(1), 61–84.

Seekings, J. (1993) Gibraltar: Developing tourism in a political impasse. *Tourism Management* 14(1), 61–67. DOI: 10.1016/0261-5177(93)90083-W.

Serzhanova, V. (2014) Concept and status of fictional states. *Annales Universitatis Apulensis Series Jurisprudentia* 17, 82–92.

Sharpley, R. (2001) Tourism in Cyprus: Challenges and opportunities. *Tourism Geographies* 3(1), 64–86. DOI: 10.1080/14616680010008711.

Sharpley, R. (2003) Tourism, modernisation and development on the island of Cyprus: challenges and policy responses. *Journal of Sustainable Tourism* 11(2-3), 246–265. DOI: 10.1080/09669580308667205.

Sharpley, R. (2012) Island tourism or tourism on islands? *Tourism Recreation Research* 37(2), 167–172. DOI: 10.1080/02508281.2012.11081701.

Sideri, E. (2012) 'The land of the golden fleece': conflict and heritage in Abkhazia. *Journal of Balkan and Near Eastern Studies* 14(2), 263–278. DOI: 10.1080/19448953.2012.681936.

Simpson, A.W. (2014) The security of the European micro-states. In: Archer, C., Bailes, A.J.K. and Wivel, A. (eds) *Small States and International Security: Europe and Beyond*. Routledge, London, pp. 167–183.

Singh, M. (2019) Tax havens: all you need to know. *Investopedia*, 25 June 2019. Available at: https://www.investopedia.com/articles/tax/09/tax-havens.asp (accessed 3 March 2020).

Singleton, P.A. (2016) The geopolitics of future tourism development in an expanding EU. *Journal of Tourism Futures* 2(2), 155–164. DOI: 10.1108/JTF-05-2015-0033.

Smídová, A. and Smíd, T. (2020) Unfinished story of the Donetsk and Luhansk People's Republics: towards a de facto state? In: Hoch, T. and Kopeček, V. (eds) *De Facto States in Eurasia*. Routledge, London, pp. 136–155.

Smith, O. (2019) Why do so few people visit the Isle of Man? The Telegraph, 20 October 2019. Available at: https://www.telegraph.co.uk/travel/destinations/europe/articles/isle-of-man-so-few-tourists/ (accessed 3 January 2020).

Souleimanov, E.A. (2020) Terminology. In: Hoch, T. and Kopeček, V. (eds) *De Facto States in Eurasia*. Routledge, London, pp. 11–15.

Souter, D. (1984) An island apart: a review of the Cyprus problem. *Third World Quarterly* 6(3), 657–674. DOI: 10.1080/01436598408419791.

Sovereign Order of Malta (2020) Order of Malta. Available at: http://www.orderofmalta.int/ (accessed 1 April 2020).

Spence, E. (2016) Performing wealth and status: observing super-yachts and the super-rich in Monaco. In: Hay, I. and Beaverstock, J.V. (eds) *Handbook on Wealth and the Super-Rich*. Elgar, Cheltenham, pp. 287–301.

Spierings, B. and van der Velde, M. (2013) Cross-border differences and unfamiliarity: shopping mobility in the Dutch-German Rhine-Waal Euroregion. *European Planning Studies* 21(1), 5–23. DOI: 10.1080/09654313.2012.716236.

Stan, L. (2018) Transitional justice: lessons from Moldova. In: Horne, C.M. and Stan, L. (eds) *Transitional Justice and the Former Soviet Union: Reviewing the Past, Looking toward the Future*. Cambridge University Press, Cambridge, pp. 263–281.

States of Guernsey (2018) *Guernsey Facts and Figures, 2018*. States of Guernsey Data and Analysis, Saint Peter Port.

States of Jersey (2020a) Jersey's relationship with the UK and the EU. Available at: https://www.gov.je/Government/Departments/JerseyWorld/pages/relationshipeuanduk.aspx#anchor-0 (accessed 13 February 2020).

States of Jersey (2020b) Tourism statistics. Available at: https://www.gov.je/Government/JerseyInFigures/BusinessEconomy/Pages/Tourism.aspx#anchor-0 (accessed 22 January 2020).

Stato della Città del Vaticano (1979) *The Vatican City State: Explanatory Notes*. Vatican Polyglot Press, Vatican City.

Stato della Città del Vaticano (2020) Popolazione. Available at: https://www.vaticanstate.va/it/stato-governo/note-generali/popolazione.html (accessed 1 April 2020).

Staudenmaier, R. (2018) Europe's overseas territories: What you need to know. *DW*, 3 November 2018. Available at: https://www.dw.com/en/europes-overseas-territories-what-you-need-to-know/a-46145298 (accessed 22 April 2020).

Steiger, R., Scott, D., Abegg, B., Pons, M. and Aall, C. (2019) A critical review of climate change risk for ski tourism. *Current Issues in Tourism* 22(11), 1343–1379.

Steinmetz, R. and Wivel, A. (eds) (2010) *Small States in Europe: Challenges and Opportunities*. Ashgate, Farnham.

Stephenson, C. (1942) *Mediaeval Feudalism*. Cornell University Press, Ithaca.

Stoffelen, A. (2018) Tourism trails as tools for cross-border integration: a best practice case study of the Vennbahn cycling route. *Annals of Tourism Research* 73, 91–102.

Stoffelen, A. and Vanneste, D. (2017) Tourism and cross-border regional development: insights in European contexts. *European Planning Studies* 25(6), 1013–1033.

Stojanovska-Stefanova, A. and Atanasoski, D. (2016) State as a subject of international law. *US-China Law Review* 13(1), 25–33.

Stringer, K.D. (2006) Pacific island microstates: pawns or players in Pacific rim diplomacy? *Diplomacy and Statecraft* 17(3), 547–577.

Stringer, K.D. (2013) *Sovereign Liechtenstein: The Soft Power Projection of a Very Small State*. Centre for Small State Studies, Reykjavik.

Suau-Sanchez, P. and Voltes-Dota, A. (2019) Drivers of airport scheduled traffic in European winter tourism areas: infrastructure, accessibility, competition and catchment area. *Journal of air transport management*. DOI: https://doi.org/10.1016/j.jairtraman.2019.101723.

Şveţ, A. (2013) Staging the Transnistrian identity within the heritage of Soviet holidays. *History and Anthropology* 24(1), 98–116.

Switzerland Tourism (2020) Büsingen. Available at: https://www.myswitzerland.com/en-us/destinations/switzerlands-german-island/ (accessed 29 March 2020).

Taillefer, F. (1991) Le paradoxe Andorran. *Revue Géographique des Pyrénées et du Sud-Ouest* 62(2), 117–138.

Tanganelli, D. and Pou, V. (2012) Commercial diplomacy in action to ensure a micro-state's international competitiveness during years of economic and political turmoil: the case of Andorra. *International Journal of Diplomacy and Economy* 1(2), 163–176.

Telegraph (2016) 10 things you didn't know about the Vatican City. *The Telegraph*, 17 June, 2016. Available at: https://www.telegraph.co.uk/travel/lists/surprising-facts-about-the-vatican-city/ (accessed 3 December 2019).

Thao, N.H. (2000) The China-Vietnam border delimitation treaty of 30 December 1999. *Boundary and Security Bulletin* 8(1), 87–90.

Thomassen, M. (2011) Who has heard of the Faroe Islands? *Culture Mandala: The Bulletin of the Centre for East-West Cultural and Economic Studies* 9(2), 82–90.

Thorhallsson, B. (2018) Studying small states: a review. *Small States & Territories* 1(1), 17–34.

Thorhallsson, B. (ed.) (2019) *Small States and Shelter Theory: Iceland's External Affairs*. Routledge, London.

Thürer, D. (1998) The perception of small states: myth and reality. In: Goetschel, L. (ed.) *Small States Inside and Outside the European Union: Interests and Policies*. Kluwer, Dortrecht, pp. 33–42.

Timothy, D.J. (1995) Political boundaries and tourism: borders as tourist attractions. *Tourism Management* 16(7), 525–532.
Timothy, D.J. (1996) Small and isolated: the politics of tourism in international exclaves. *Acta Turistica* 8(2), 99–115.
Timothy, D.J. (1998a) Collecting places: geodetic lines in tourist space. *Journal of Travel and Tourism Marketing* 7(4), 123–129.
Timothy, D.J. (1998b) Tourism development in a small, isolated community: the case of Northwest Angle, Minnesota. *Small Town* 29(1), 20–23.
Timothy, D.J. (2000) Borderlands: an unlikely tourist destination? *Boundary and Security Bulletin* 8(1), 57–65.
Timothy, D.J. (2001a) Benefits and costs of smallness and peripheral location in tourism: Saint-Pierre et Miquelon (France). *Tourism Recreation Research* 26(3), 63–70.
Timothy, D.J. (2001b) Postage stamps, microstates and tourism. *Tourism Recreation Research* 26(3), 85–88.
Timothy, D.J. (2001c) *Tourism and Political Boundaries*. Routledge, London.
Timothy, D.J. (2003) Where on earth is this place? The potential of non-nations as tourist destinations. *Tourism Recreation Research* 28(1), 93–96.
Timothy, D.J. (2005) *Shopping Tourism, Retailing and Leisure*. Channel View Publications, Bristol.
Timothy, D.J. (2010) Contested place and the legitimization of sovereignty claims through tourism in polar regions. In: Hall, C.M. and Saarinen, J. (eds) *Tourism and Change in Polar Regions: Climate, Environment and Experience*. Routledge, London, pp. 288–300.
Timothy, D.J. (2011) *Cultural Heritage and Tourism: An Introduction*. Channel View Publications, Bristol.
Timothy, D.J. (2017a) Andorra. In: Lowry, L.L. (ed.) *The SAGE International Encyclopedia of Travel and Tourism*. SAGE, Thousand Oaks, CA, pp. 69–70.
Timothy, D.J. (2017b) Customs and immigration. In: Lowry, L.L. (ed.) *The SAGE International Encyclopedia of Travel and Tourism*. SAGE, Thousand Oaks, CA, pp. 338–341.
Timothy, D.J. (2017c) Liechtenstein. In: Lowry, L.L. (ed.) *The SAGE International Encyclopedia of Travel and Tourism*. SAGE, Thousand Oaks, CA, pp. 755–756.
Timothy, D.J. (2017d) Monaco. In: Lowry, L.L. (ed.) *The SAGE International Encyclopedia of Travel and Tourism*. SAGE, Thousand Oaks, CA, pp. 838–840.
Timothy, D.J. (2017e) San Marino. In: Lowry, L.L. (ed.) *The SAGE International Encyclopedia of Travel and Tourism*. SAGE, Thousand Oaks, CA, p. 1070.
Timothy, D.J. (2017f) Tourism and geopolitics in the GCC region. In: Stephenson, M.L. and al-Hamarneh, A. (eds) *International Tourism Development and the Gulf Cooperation Council States: Challenges and Opportunities*. Routledge, London, pp. 45–60.
Timothy, D.J. (2018a) Geography: the substance of tourism. *Tourism Geographies* 20(1), 166–169.
Timothy, D.J. (2018b) Globalization, supranationalism and tourism. In: Cooper, C., Volo, S., Gartner, W.C. and Scott, N. (eds) *The Sage Handbook of Tourism Management, Vol 1*. Sage, London, pp. 569–581.
Timothy, D.J. (2019) Supranationalism and tourism: free trade, customs unions, and single markets in an era of geopolitical change. In: Timothy, D.J. (ed.) *Handbook of Globalisation and Tourism*. Edward Elgar, Cheltenham, pp. 100–113.
Timothy, D.J. (2020) Domestic tourism: challenging the notion through a geopolitical lens. *Tourism Review International* 24(1), 67–73.
Timothy, D.J. (in press) Tourism, mobility and transport: concerns and developments in Pacific island states. In: Stephenson, M.L. (ed.) *Routledge Handbook on Tourism in Pacific Islands States*. Routledge, London.
Timothy, D.J. and Boyd, S.W. (2015) *Tourism and Trails: Cultural, Ecological and Management Issues*. Channel View Publications, Bristol.
Timothy, D.J. and Canally, C. (2008) The role of the US-Mexico border as a destination: student traveler perceptions. *Tourism Analysis* 13(3), 259–269.
Timothy, D.J. and Gelbman, A. (2015) Tourist lodging, spatial relations, and the cultural heritage of borderlands. *Journal of Heritage Tourism* 10(2), 202–212.
Timothy, D.J., Guia, J. and Berthet, N. (2014) Tourism as a catalyst for changing boundaries and territorial sovereignty at an international border. *Current Issues in Tourism* 17(1), 21–27.
Timothy, D.J. and Ioannides, D. (2002) Tour operator hegemony: dependency, oligopoly, and sustainability in insular destinations. In: Apostolopoulos, Y. and Gayle, D.J. (eds) *Island Tourism and Sustainable Development: Caribbean, Pacific, and Mediterranean Experiences*. Praeger, Westport, CT, pp. 181–198.

Timothy, D.J. and Michalkó, G. (2016) European trends in spatial mobility. *Hungarian Geographical Bulletin* 65(4), 317–320.

Timothy, D.J. and Olsen, D.H. (2001) Challenges and opportunities of marginality in the Arctic: a case of tourism in Greenland. *Tourism* 49(4), 299–308.

Timothy, D.J. and Olsen, D.H. (2018) Religious routes, pilgrim trails: spiritual pathways as tourism resources. In: Butler, R. and Suntikul, W. (eds) *Tourism and Religion: Issues and Implications*. Channel View Publications, Bristol, pp. 220–235.

Timothy, D.J., Prideaux, B. and Kim, S.S. (2004) Tourism at borders of conflict and (de)militarized zones. In: Singh, T.V. (ed.) *New Horizons in Tourism: Strange Experiences and Stranger Practices*. CABI, Wallingford, pp. 83–94.

Timothy, D.J. and Saarinen, J. (2013) Cross-Border co-operation and tourism in Europe. In: Costa, C., Panyik, E. and Buhalis, D. (eds) *Trends in European Tourism Planning and Organisation*. Channel View Publications, Bristol, pp. 64–74.

Timothy, D.J., Saarinen, J. and Viken, A. (2016) Tourism issues and international borders in the Nordic region. *Scandinavian Journal of Hospitality and Tourism* 16(1), 1–13.

Toal, G. and O'Loughlin, J. (2013) Land for peace in Nagorny Karabakh? Political geographies and public attitudes inside a contested de facto state. *Territory, Politics, Governance* 1(2), 158–182.

Togan, S. (2015) The EU-Turkey Customs Union: a model for future Euro-Med integration. In: Ayadi, M. Dabrowski. and De Wulf, L. (eds) *Economic and Social Development of the Southern and Eastern Mediterranean Countries*. Springer, Cham, Switzerland, pp. 37–48.

Toschi, U. (1931) The Vatican City state from the standpoint of political geography. *Geographical Review* 21(4), 529–538.

TravelMole (2020) Arrivals from Channel Islands will be exempt from UK quarantine rule. TravelMole, 12 May 2020. Available at: https://www.travelmole.com/news_feature.php?c=setreg&region=2&m_id=Y!T__rbT_db&w_id=37519&news_id=2042646 (accessed 13 May 2020).

Troitiño, D.R., Kerikmäe, T., Chochia, A. and Hrebickova, A. (2018) First European and Pan-European integration efforts and British reluctance. In: Troitiño, D.R., Kerikmäe, T. and Chochia, A. (eds) *Brexit: History, Reasoning and Perspectives*. Springer, Cham, Switzerland, pp. 3–20.

UNESCO (2020) World Heritage List. Available at: https://whc.unesco.org/en/list/ (accessed 30 January 2020).

United Nations (2020) Per capita GDP at current prices. Available at: https://data.un.org/Data.aspx?d=SNAAMA&f=grID%3A101%3BcurrID%3AUSD%3BpcFlag%3A1 (accessed 10 March 2020).

US Department of State (1987) *The Holy See*. Bureau of Public Affairs, US Department of State, Washington, DC.

Vaccaro, I. and Beltran, O. (eds) (2010) *Social and Ecological History of the Pyrenees: State, Market and Landscape*. Left Coast Press, Walnut Creek, CA.

Vaello, O. (2018) Visit a country that doesn't exist: here's why bold travelers should visit Transnistria, a mysterious land in limbo. *National Geographic*, 11 January 2018. Available at: https://www.nationalgeographic.com/travel/destinations/europe/things-to-see-transnistria/ (accessed 3 November 2019).

van Bets, L.K., Lamers, M.A. and van Tatenhove, J.P. (2017) Collective self-governance in a marine community: expedition cruise tourism at Svalbard. *Journal of Sustainable Tourism* 25(11), 1583–1599.

van Geluwe, T. (2003) *Baarle: A Unique Village in the World*. Province Antwerpen, Baarle-Hertog.

Vanegas, M.V. and Croes, R.R. (2003) Growth, development and tourism in a small economy: evidence from Aruba. *International Journal of Tourism Research* 5(5), 315–330.

Vatican Secretary of State (2020) Liste du Corps Diplomatique Près le Saint-Siège. Available at: https://www.sdsofficium.va/cd/corpsdiplomatique.html (accessed 1 April 2020).

Väylävirasto (2020) The Saimaa Canal. Available at: https://vayla.fi/web/en/waterways/canals-and-bridges/the-saimaa-canal (accessed 3 May 2020).

Veenendaal, W. (2013a) Political representation in microstates: St. Kitts and Nevis, Seychelles, and Palau. *Comparative Politics* 45(4), 437–456.

Veenendaal, W. (2013b) Size and personalistic politics: characteristics of political competition in four microstates. *The Round Table* 102(3), 245–257.

Veenendaal, W.P. (2015a) Democracy in microstates: why smallness does not produce a democratic political system. *Democratization* 22(1), 92–112.

Veenendaal, W.P. (2015b) *Politics and Democracy in Microstates*. Routledge, London.

Veenendaal, W.P. (2016a) Analyzing the foreign policy of microstates: the relevance of the international patron-client model. *Foreign Policy Analysis* 13(3), 561–577.
Veenendaal, W.P. (2016b) Monarchy and democracy in small states: an ambiguous symbiosis. In: Wolf, S. (ed.) *State Size Matters: Politik und Recht im Kontext von Kleinstaatlichkeit und Monarchie*. Springer, Wiesbaden, pp. 183–198.
Veenendaal, W.P. (2020) Politics of the four European microstates: Andorra, Liechtenstein, Monaco and San Marino. In: Baldacchino, G. and Wivel, A. (eds) *Handbook on the Politics of Small States*. Edward Elgar, Cheltenham, pp. 150–167.
Veenendaal, W. and Wolf, S. (2016) Concluding remarks: achievements, challenges, and opportunities of small state research. In: Wolf, S. (ed.) *State Size Matters: Politik und Recht im Kontext von Kleinstaatlichkeit und Monarchie*. Springer, Wiesbaden, pp. 277–284.
Viken, A. (2011) Tourism, research, and governance on Svalbard: a symbiotic relationship. *The Polar Record* 47(4), 335–347.
Vinokurov, E. (2007) *A Theory of Enclaves*. Lexington Books, Lanham, MD.
Vital, D. (2006) The inequality of states: a study of the small power in international relations. In: Ingebritsen, C., Neumann, I.B. and Gstöhl, S. (eds) *Small States in International Relations*. University of Washington Press, Seattle, pp. 77–88.
Vitaliev, V. (2008) *Passport to Enclavia: Travels in Search of a European Identity*. Reportage Press, London.
Vitaliev, V. (2016) 9 European 'enclaves'—the continent's quirkiest. *The Telegraph*, August 7. Available at: https://www.telegraph.co.uk/travel/destinations/europe/articles/quirky-european-enclaves-where-cultures-collide/ (accessed 30 December 2019).
Vitaliev, V. (2019) After all: Railways-A historic German transport link that remains forever Belgian. *Engineering & Technology* 14(6), 98.
Vítová, P., Harmáček, J. and Opršal, Z. (2019) Determinants of tourism flows in small island developing states (SIDS). *Island Studies Journal* 14(2), 3–22.
Volgger, M., Pechlaner, H. and Hoch, M. (2015) Liechtenstein, tourism. In: Jafari, J. and Xiao, H. (eds) *Encyclopedia of Tourism*. Springer, Cham, Switzerland, pp. 555–557.
Vukonić, B. (2006) Sacred places and tourism in the Roman Catholic tradition. In: Timothy, D.J. and Olsen, D.H. (eds) *Tourism, Religion and Spiritual Journeys*. Routledge, London, pp. 237–253.
Waldman, C. and Mason, C. (2006) *Encyclopedia of European Peoples*. Facts on File, New York.
Weaver, D. (2017) Core–periphery relationships and the sustainability paradox of small island tourism. *Tourism Recreation Research* 42(1), 11–21.
Weaver, D. (2018) Creative periphery syndrome? Opportunities for sustainable tourism innovation in Timor-Leste, an early stage destination. *Tourism Recreation Research* 43(1), 118–128.
Webster, C. and Timothy, D.J. (2006) Travelling to the 'other side': the occupied zone and Greek Cypriot views of crossing the Green Line. *Tourism Geographies* 8(2), 162–181.
Wettenhall, R. (2018) A journey though small state governance. *Small States & Territories* 1(1), 111–128.
Whyte, B.R. (2002) Bordering on the ridiculous? A comparison of the Baarle and Cooch Behar enclaves. *The Globe* 53, 43–61.
Whyte, B.R. (2004) *En Territoire Belge et Quarante Centimètres de la Frontière": An Historical and Documentary Study of the Belgian and Dutch Enclaves of Baarle-Hertog and Baarle-Nassau*. School of Anthropology, Geography and Environmental Studies, University of Melbourne, Melbourne.
Więckowski, M. (2005) The contemporary Polish-Slovak transboundary cooperation. In: Potrykowska, A. (ed.) *Population, Environment and Development*. Polish Academy of Sciences, Warsaw, pp. 187–200.
Więckowski, M. (2010) Tourism development in the borderlands of Poland. *Geographia Polonica* 83(2), 67–81.
Więckowski, M. and Cerić, D. (2016) Evolving tourism on the Baltic Sea coast: perspectives on change in the Polish maritime borderland. *Scandinavian Journal of Hospitality and Tourism* 16(1), 98–111.
Wilkinson, P.F. (1989) Strategies for tourism in island microstates. *Annals of Tourism Research* 16(2), 153–177.
Wilkinson, P.F. (1994) Tourism policy and planning in the eastern Caribbean: four case studies. *Tourism Recreation Research* 19(2), 5–17.
Wilson, G.N., Johnson, H. and Sallabank, J. (2015) 'I'm not dead yet': a comparative study of indigenous language revitalization in the Isle of Man, Jersey and Guernsey. *Current Issues in Language Planning* 16(3), 259–278.
Wivel, A. (2020) Small states in Europe. In: Baldacchino, G. and Wivel, A. (eds) *Handbook on the Politics of Small States*. Edward Elgar, Cheltenham, pp. 99–112.

Wivel, A. and Oest, K.J.N. (2010) Security, profit or shadow of the past? Explaining the security strategies of microstates. *Cambridge Review of International Affairs* 23(3), 429–453. DOI: 10.1080/09557571.2010.484047.

Wolf, S. (2015) Different approaches, different results in small state studies: complementary views on the monarchy and traditional governance in Liechtenstein. *Swiss Political Science Review* 21(2), 350–361. DOI: 10.1111/spsr.12151.

Wolf, S. (2016a) Die Erforschung von Politik und Recht in Kleinstaat und Monarchie – Eine konzeptionelle Einführung. In: Wolf, S. (ed.) *State Size Matters: Politik und Recht im Kontext von Kleinstaatlichkeit und Monarchie*. Springer, Wiesbaden, pp. 1–12.

Wolf, S. (2016b) Governance in small state legislatures: a theoretical framework and the case of Liechtenstein. In: Wolf, S. (ed.) *State Size Matters: Politik und Recht im Kontext von Kleinstaatlichkeit und Monarchie*. Springer, Wiesbaden, pp. 83–97.

Wolf, S., Bussjäger, P. and Schiess Rütimann, P.M. (2018) Law, small state theory and the case of Liechtenstein. *Small States & Territories* 1(2), 183–196.

World Prison Brief (2020) San Marino. Available at: https://www.prisonstudies.org/country/san-marino (accessed 30 March 2020).

World Tourism Organization (2008) *Yearbook of Tourism Statistics, Data 2002-2006*. UNWTO, Madrid.

World Tourism Organization (2010) *Yearbook of Tourism Statistics, Data 2004-2008*. UNWTO, Madrid.

World Tourism Organization (2011) *Compendium of Tourism Statistics, Data 2005-2009*. UNWTO, Madrid.

World Tourism Organization (2014) *Yearbook of Tourism Statistics, Data 2008-2012*. UNWTO, Madrid.

World Tourism Organization (2015) *Compendium of Tourism Statistics, Data 2009-2013*. UNWTO, Madrid.

World Tourism Organization (2020a) *Compendium of Tourism Statistics, Data 2014-2018*. UNWTO, Madrid.

World Tourism Organization (2020b) e-Library. Available at: https://www.e-unwto.org/ (accessed 3 April 2020).

World Tourism Organization (2020c) Growth in international tourist arrivals continues to outpace the economy. *World Tourism Barometer* 18(1), 1–6.

World Tourism Organization (2020d) *Yearbook of Tourism Statistics, 2014-2018*. UNWTO, Madrid.

Wróblewski, Ł. and Kasperek, A. (2019) Euroregion as an entity stimulating the sustainable development of the cross-border market for cultural services in a city divided by a border. *Sustainability* 11(8), 2232–21. DOI: 10.3390/su11082232.

Yasarata, M., Altinay, L., Burns, P. and Okumus, F. (2010) Politics and sustainable tourism development–Can they co-exist? Voices from North Cyprus. *Tourism Management* 31(3), 345–356. DOI: 10.1016/j.tourman.2009.03.016.

Yoon, M.Y. (2014) Colonialism and border disputes in Africa. *The Journal of Territorial and Maritime Studies* 1(1), 75–89.

Zhang, C. (2019) Mobile borders and turbulent mobilities: mapping the geopolitics of the Channel Tunnel. *Geopolitics* 24(3), 728–755. DOI: 10.1080/14650045.2017.1379994.

Zonneveld, W. and Faludi, A. (1997) Vanishing borders: the second Benelux structural outline. *Built Environment* 23(1), 5–13.

Index

Page numbers in *italics* refer to illustrations.